DISCARD

The Isaac Newton
School of Driving

PUBLISHING FOR THE WORLD
125 Years

THE JOHNS HOPKINS UNIVERSITY PRESS

THE ISAAC NEWTON

SCHOOL OF DRIVING
Physics & Your Car

Barry Parker

The Johns Hopkins University Press
Baltimore & London

© 2003 The Johns Hopkins University Press
All rights reserved. Published 2003
Printed in the United States of America on acid-free paper
9 8 7 6 5 4 3 2 1

The Johns Hopkins University Press
2715 North Charles Street
Baltimore, Maryland 21218-4363
www.press.jhu.edu

Library of Congress Cataloging-in-Publication Data

Parker, Barry, 1935–
    The Isaac Newton school of driving : physics and your car /
Barry Parker.
        p. cm.
    Includes bibliographical references and index.
    ISBN 0-8018-7417-3 (alk. paper)
        1. Mechanics. 2. Elasticity. 3. Motion. 4. Thermodynamics.
    I. Title: Physics and your car. II. Title.

QC125.2.P37 2003
531—dc21   2003047527

A catalog record for this book is available from the British
Library.

Figures 5–89, 96–101 by Lori Scoffield-Beer

To Charles and Olive Vizer

Contents

Acknowledgments

I am grateful to Trevor Lipscombe for his many suggestions and help in preparing this volume. Furthermore, I would like to thank Pat Negyesi for his help in obtaining photographs. And finally, I thank Alice Calaprice for her careful editing of the manuscript, and the staff of the Johns Hopkins University Press for their assistance in bringing this project to completion.

The Isaac Newton
School of Driving

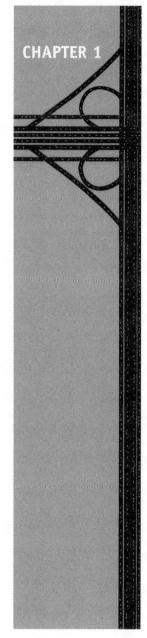

CHAPTER 1

Introduction

New cars, with their sleek, shiny, curved lines, are objects of intrigue, elegance, grace, and beauty. They are exciting, and they can be a lot of fun. The thrill of the first drive in your new car is something you remember for a long time. The science of physics also has a certain beauty and elegance. With a few simple principles and the enormous power of mathematics, you can make amazingly accurate predictions about everything from atomic interactions to the expansion of the universe. And you can also make important predictions about cars.

In this book I bring cars and physics together. At first you might not think they have a lot in common, but they do. It's easy to show that every branch of physics is represented somewhere within an automobile. Mechanics, the branch of physics that deals with motion, is particularly applicable. After all, cars move, and if they are moving they have a certain velocity, and to set them in motion you have to accelerate them. Furthermore, to accelerate them you have to apply a force, and this force comes from a source of energy. Physics is at the basis of all of this. Indeed, two of the main terms that are used in relation to cars are *horsepower* and *torque,* both of which are important terms in mechanics.

Another important branch of physics deals with elasticity and vibrational motion, and these concepts are significant in relation to a car's suspension system. Heat and thermodynamics are critical in engine performance, while electricity and magnetism are what allow you to start the engine and keep it running. With modern additions such as telematics (see chapter 11), wireless communication—which takes place by means of electromagnetic waves—is becoming increasingly significant in cars. It's easy to see that physics is critical to the understanding of cars; it also has helped improve them and has made them much safer.

I taught physics at the university level for 30 years and I grew up around cars, so it's perhaps natural that the two eventually came together for me. My father was a mechanic and a garage owner. As a teenager, I worked in everything from the parts department to the collision repair and lubrication departments. I don't think anyone would have trusted me to do a full-scale repair on an engine at that time, but that didn't stop me from taking apart the motor in my own car. For a while I thought of becoming a mechanical engineer and designing cars. It was something that particularly appealed to me, but after looking into it I realized there were not a lot of jobs in the area so I ended up going into physics. After I graduated and began teaching I soon realized that my students had a lot of interest in cars. Whenever I used examples from the "automotive world" to illustrate a principle of physics, their interest seemed to pique. So I continued the practice as much as possible.

One of the things I enjoyed when I worked at the garage was driving the new cars. A car that stands out in my memory is a yellow convertible. We didn't see many convertibles at that time, so this one was a novelty. After referring to it as "yellow" several times, I was politely informed that it wasn't yellow—it was "sportsman's green." It didn't look green to me, however, and I continued to think of it as yellow. In any case, much to my delight, I managed to drive it around town a few times. What made me think about the car now was a recent issue of *Automobile* magazine, which featured the Lamborghini Murciélago on its cover. Now, *that's* what I call "sportman's green." You've heard of shocking pink; well, this one is "shocking green." The article about the car is appropriately titled "Bat Out of Hell." Upon reading it I learned that the Lamborghini gets its name from a Spanish fighting bull. Apparently the first Murciélago's life was spared in the bullring by a famous matador in 1879, who admired its fighting spirit and courage. Instead of killing it, he named a car for it (though that happened much later).

The Lamborghini is a beautiful car (fig. 1). In fact, everything is beautiful about it except the price, which at $273,000 is out of my range. Since I'm writing a book on cars you might wonder what I drive. As an avid fisherman, hiker, and skier, I've owned mostly SUVs over the past few years. They seem to fit in best with my lifestyle.

Unlike most science books, this book does not get more complex as you get into it; in fact, chapter 2 might be the most complicated part of it, since it contains more mathematics than most of the later chapters. I've tried to limit the math, but a certain amount is needed for a good understanding of the physics. In some cases I've omitted the derivation of a formula, but this shouldn't affect the understanding.

Chapter 2 deals with the basic physics of driving. It's concerned with speed, velocity, acceleration, and the forces you experience when you're in a car. The level of the physics is approximately that of a high school physics class, and concepts such as momentum, energy, inertia, centripetal force, and torque are explained. All of these aspects are important in relation to the motion of a car.

Fig. 1. 2002 Lamborghini
Murciélago (Lamborghini).

Chapter 3 is a key chapter. It's about engines and how they work, and that, of course, is what cars are all about. You wouldn't get very far without an engine. I start out with a little history, which is interesting to most people, but the central topic is the four-stroke combustion engine and how it works. Efficiency is a critical part of any engine and I define several different efficiencies: mechanical efficiency, combustion efficiency, thermal efficiency, and volumetric efficiency. Everyone is interested in comparing not only the efficiency but the horsepower, torque, and other elements of modern cars, so I include several tables for that purpose. I also discuss turbochargers and superchargers briefly, and the role of heat in the engine. The chapter ends with a discussion of diesel engines and the rotary engine.

Chapter 4 is about the electrical system of the car, and I've had many interesting experiences with electrical systems. One of the earliest occurred many years ago, before I knew much about them. My wife and I were coming home from a short vacation. It was dark and the rain was pouring down so hard that the wipers could hardly keep up with it. All of a sudden the engine stopped. I wasn't sure what to do; I didn't even have a flashlight in the car, but I knew I had to check under the hood. After all, something was wrong, and it was possible it could be fixed. So out I went and popped up the hood. The only light I had was from passing cars but I quickly checked everything I could. There were no loose wires, and I checked the points, condenser, and rotor. As far as I could see, everything was okay, but by the time I was finished I was soaked through to the skin. I put the hood down and got back in the car.

"Did you get it fixed?" my wife asked.

I shrugged as I tried the starter, and as I expected it didn't start. To make a long story short, I sat for another fifteen minutes, not sure what to do, when to my relief it suddenly started. I found out later it was a problem with the electrical system. I decided then and there that I better learn more about this system.

So in chapter 4, I cover the basics of electricity and circuits, but I also talk about the starter and the basic physics behind it, and the alternator or generator and how they are regulated. Finally, I discuss the ignition system and the interesting physics behind it.

Chapter 5 is about brakes, and brakes are all about friction, which is an important element throughout physics. Brakes are one of the most important parts of a car in the sense that if they aren't working properly, or the braking conditions are poor, you're in trouble. When I worked in the garage, one of my jobs was to jockey cars back and forth between some of the nearby towns. One morning I was told to take an old half-ton pickup to the next town, so I looked it over as I got ready to go. The hood didn't seem to be snapped down properly so I banged on it a couple of times, but that didn't help much; nevertheless, I didn't worry about it. The road to the next town was along a lake, and I knew the dropoff in the lake was steep—the water was black only a few feet from the edge, and there was no sign of the bottom. I had never given the dropoff much thought before, however.

Fig. 2. 2002 Dodge Viper (Chrysler).

As I got the truck up to about 60 mph I noticed that the hood was starting to vibrate. Soon it was banging. I didn't like the sound of it, but since I had checked it before I left I was sure nothing would happen.

Suddenly something crashed into the windshield. It made such an enormous noise that I'm sure I jumped a few inches off the seat. I was so startled that it took me a moment to realize what had happened: the hood had flown up across the windshield. I couldn't see a thing. All I could think about was the dropoff that was only a few feet away from the edge of the road.

I tried to roll down the window as quickly as possible, but it only rolled partway. By now I had jammed on the brakes, but I was still doing 40 or 50. If I didn't get it stopped fast I knew where I would end up. I opened the door and peered out and was surprised to see that I was still on the road—I had expected to hit the water any second. Somehow, I finally managed to come to a stop.

There wasn't much that was good about the old truck, but—thank heavens—it had good brakes. And that experience gave me an appreciation of brakes.

In chapter 5 we will also look at the different types of friction, stopping distances, tire traction, hydraulics and the brake system, and ABS, the antilock braking system.

From brakes we turn to the suspension system and transmission in chapter 6. Although they aren't related to each other, I cover both in the same chapter. Both have a lot of physics associated with them. In Giancarlo Genta's comprehensive book *Motor Vehicle Dynamics* the chapter on suspension systems is one of the longest and most complicated in the book. He deals with every possible aspect of suspension systems in considerable mathematical detail, and unless you have a degree in engineering it's unlikely you could understand it. It's not my intent to get into that much detail, but I would like to give you some feeling for the basics of suspension systems, and the physics behind them.

The transmission system is, in many ways, the most complicated system in the car. It is the major component of the power train to the back wheels. The power train has the task of conveying the rotary motion developed in the engine to the back wheels, and overall there are a lot of things that have to be taken into consideration. The basic component of the transmission system is the circular, toothed gear, which itself is quite simple. The complication comes when you start bringing gears together. In this chapter we will look at how gears work and why planetary gears and compound planetary gears are needed in a car.

In chapter 7 we look at the aerodynamics of cars. I've always had an interest in aerodynamics. I suppose it goes back to my interest in airplanes when I was young: model building was my passion, and I spent much of my

Fig. 3. 2002 Chevy
Corvette White Shark
(General Motors).

time building and flying model airplanes. Central to this
chapter is the coefficient of aerodynamic drag, which
tells us virtually everything there is to know about the
aerodynamics of cars. As we will see, this coefficient
has slowly been coming down over the years. In other
words, cars have become more aerodynamic, and not
only does this give the car more eye appeal, but it saves
a lot of gas. In this chapter, we will also look at drag
forces of all types, streamlines and airflow around a car,
Bernoulli's theorem, and aerodynamic lift and downforce and how they affect the stability of the car.

Chapter 8 is a brief course on car collisions. Physics
is, after all, the science of objects that interact with each
other: atoms smashing into one another, molecules of
a gas colliding with one another, and balls and other
objects of various kinds bouncing off one another. It's

pretty obvious that this concept can be extended to the collisions of cars. I'm not sure it will help you avoid a collision, but it will give you an appreciation of the tremendous forces involved in a crash, which may give you an additional incentive to avoid one. So, in this chapter we will discuss head-on collisions, glancing collisions, the reconstruction of accidents to determine how fast the cars were going, and crash tests.

In chapter 9 we turn to the physics of auto racing. Auto racing is, indeed, a popular sport, and it has a lot of fans. I just finished watching the biography of Enzo Ferrari on TV, and I was intrigued. It was the fascinating story of someone who overcame many difficulties and ended up a legend in the annals of car racing. In 1916, while still a teenager, he lost both his father and his brother, then two years later he almost died of influenza. With the end of World War I, he was penniless and couldn't find a job. Still, something happened at this time that changed his life. He tried to get a job at Fiat, the largest car manufacturer in Italy, and was turned down. At the time, Fiat had the best racing cars in the world, and Ferrari vowed he would build race cars that were even better than Fiat's—and indeed he did. What I found particularly interesting was that Ferrari, who is now renowned for the Ferrari luxury cars, had almost no interest in any of the cars he manufactured except the racing cars. Racing was his passion, and he made luxury cars only to get money for his racing cars. His drivers won over five thousand races during his lifetime, a record for any one person. Nevertheless, he will no doubt be most remembered for his luxury cars.

There's a lot of physics in car racing. Tires, the shifting of weight as the car moves, the position of the center of gravity of the car, the moment of inertia of the car—all are important, and all are determined by physics. Racing strategy, which is critical to the racing driver, also depends on the principles of physics.

Chapter 10 is a little different from the preceding chapters. It's on traffic, or more specifically, on traffic congestion. Everyone has encountered traffic congestion at one time or another, and I always breathe a little easier when I finally get through a large city's traffic jams. Luckily, I don't have to battle large amounts of traffic every day as some people do, and I feel for them. I looked forward to writing this chapter because it involved one of my favorite topics: chaos. I had, in fact, just written a book on chaos, so that was helpful.

Some of my books were displayed at a recent book fair. One of them was the book on chaos. Someone picked it up and looked carefully at the colorful picture on the cover.

Fig. 4. 2002 Ford Thunderbird (Ford Motor Co.).

"H'm . . . chaos," he said. "Sounds interesting. What is it?"

I was sure the technical definition ("sensitive dependence on initial conditions") wouldn't appeal to him, so I explained that an example of chaos is the path of a leaf floating on a turbulent stream that was filled with rapids.

"Oh," he said, with a puzzled look on his face. "Why would anyone want to study that?"

Even though I didn't say it, I wanted to tell him that chaos was revolutionizing all of the sciences and thereby changing the world.

Chaos is, indeed, an intriguing branch of physics (if I may call it that), and it is growing in importance. One of its major applications in recent years has been to study traffic congestion, with some important results. Another closely allied area, referred to as *complexity,* is also giving insights into traffic control. Complexity is concerned with complex phenomena that are not quite chaotic.

Chapter 11 is about cars of the future and some of the devices they are likely to contain. Whenever I think of cars of the future I'm reminded of the TV show *Knight Rider,* which ran from 1984 to 1986. It starred David Hasselhoff and his talking car KITT. KITT was a futuristic TransAm with a quirky personality which had turbo boosters and helped to solve crimes. Of course, with such a hero car, there also had to be a villain car, and it was called KATT. The show made Hasselhoff and KITT household names for several years, and fan clubs still exist. In this chapter we look at electric hybrids, fuel cells, flywheels, and ultracapacitors, along with telematics and all the innovations it is likely to bring.

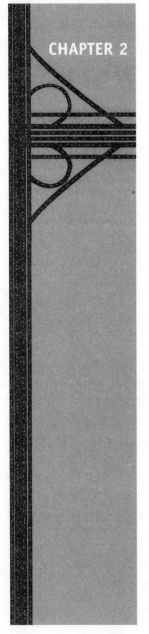

CHAPTER 2

The Open Road

Basic Physics of Driving

Imagine that you have just visited your local car dealer. You were set on either a Lamborghini or a Porsche Turbo, but when you saw the shiny red Ferrari Spider with its sleek aerodynamic lines you instantly fell in love with it. The salesperson told you it would go from 0 to 60 miles per hour in less than 4 seconds, and you couldn't wait to test drive it. As you negotiate the curves on the winding coastal mountain road, you breathe deeply, savoring the fresh ocean air. Glancing down at the ocean below you see waves hurling themselves onto the rocks. The Ferrari Spider is everything you had hoped. It seems to have horsepower to spare. The salesperson had told you that the horsepower was 400 at 8500 rpm (revolutions per minute) and the car had a maximum torque of 380 lb-ft at 4700 rpm, but he didn't explain the significance of these numbers.

Physicists, like potential car buyers, also take a great interest in acceleration, power, and aerodynamic drag. These quantities have precise technical meanings in physics, ones that turn out to be very useful in studying not only cars but also activities as diverse as baseball, golf, bicycle racing, and bird flight. Take velocity, for example. You may think of it as the speed of the vehicle, but to the physicist it's more than that. Velocity is speed with a direction attached to it; in other words, it's 60 miles per hour along Route 66 going toward Chicago. Because it has both size and direction it is called a *vector*; speed is what is called a *scalar* (it has no direction). Another example of a scalar is temperature (like speed, it has only magnitude).

When you push down on the accelerator, you change the speed—in other words, you accelerate. Actually, you accelerate if you change either your speed or your direction. So, even if you're traveling uniformly at 25 miles per hour while rounding a corner, you are accelerating. That means you are accelerating and decelerating most of the time when you take a trip through the busy streets of a city, either by stepping on the gas, braking, or turning the steering wheel.

Something else is important in the way velocity and acceleration are determined. By definition, velocity is the distance traveled in a certain time, and acceleration is the change in velocity in a certain time. If the time interval is very long, we are actually calculating the average velocity and average acceleration over that time. For example, if my journey home is 10.6 miles and it takes 32 minutes, my average speed is 19.9 miles per hour. But what we really want are instantaneous values. They are obtained by making the time interval very short. The speed shown on the speedometer in your car, for example, is an instantaneous value.

Since we'll be referring to velocities and accelerations frequently in the upcoming chapters, it's a good idea to consider their units. The unit of acceleration, in particular, can be confusing. Velocity is given as so many miles per hour (mph), or feet per sec (ft/sec) (in Europe and Canada they use meters per second or kilometers per hour). Since acceleration is (change in velocity)/time, an acceleration of, say, 15 will have units of 15 (ft/sec)/sec or more briefly 15 ft/sec². Another unit that is commonly used when dealing with acceleration is "gee's" or just "g's." You have likely heard about it in relation to rockets and space travel, but it is also important on the racetrack. Race drivers are always conscious of the "g's" they are experiencing. One "g" is 32ft/sec², and it's about the maximum your tires can withstand.

Engineers find it useful to convert between miles per hour and feet per sec. There are 5280 feet in a mile, and 3600 seconds in an hour, so one mile per hour is 5280 feet in 3600 seconds. The conversion factor is therefore 5280/3600 = 22/15, so a speed such as 60 mph is 60 × 22/15 = 88 ft/sec.

Since acceleration a is the change of velocity v in time t, then speaking very loosely we can say that velocity = acceleration × time. In the shorthand of mathematics, this is

$$v = at,$$

where v is velocity, a is acceleration, and t is time. This is a useful formula: if we know what our acceleration is, and how long we accelerate, we can predict our final speed. If the acceleration is, for example, 10 ft/sec², our velocity at the end of 10 seconds would be 100 ft/sec, or converting to miles per hour, it is 15/22 × 100 = 68 mph.

How far do we travel while accelerating? Let's say we accelerate at 100 ft/sec². How far do we go in 10 seconds? To determine this, let's begin with average velocity. We get it by dividing the sum of the initial and final velocities by two. But the initial velocity is zero, so we have average velocity = (0 + at)/2 or at/2. The distance traveled during this time is therefore

$$d = vt = \tfrac{1}{2}at \times t = \tfrac{1}{2}at^2.$$

If our acceleration is 100 ft/sec^2 we will travel ½ × 100 × 10^2 = 5000 feet in 10 seconds.

We can make a plot of the distance traveled during various times for a given acceleration. Let's assume three different accelerations of 50 ft/sec^2, 100 ft/sec^2, and 150 ft/sec^2. We get a curved line—a parabola (fig. 5). At the end of one second we see that for an acceleration of 100 ft/sec^2 the car has traveled only 50 feet, but by the end of 4 seconds it has traveled 800 feet—quite a difference.

Zero to Sixty

One of the measures of a car's power (and, to many people, how "cool" the car is) is its ability to accelerate—in particular, its ability to accelerate from 0 to 60 mph. Table 1 gives some representative times for cars and SUVs for the year 2002.

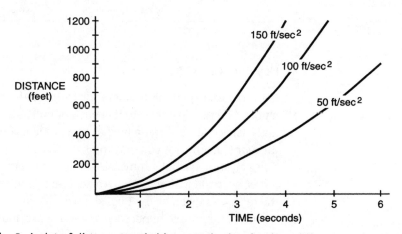

Fig. 5. A plot of distance traveled in a certain time for three different accelerations.

Table 1. Time from 0 to 60 in seconds for several models of 2002 cars

Type of Vehicle	Model	0 to 60 (sec)
Family sedans	Ford Focus ZTS	9.6
	Dodge Stratus ES	8.5
	Honda Accord EX V-6	7.6
	Hyundai XG300	8.9
Sports sedans	Ford Thunderbird	7.0
	Jaguar XK8	6.7
	Lexus SC430	5.9
	Porsche 911 GT2	4.1
	Chevy Corvette Z06	4.0
	Audi A6	6.7
	BMW 540i	6.6
	Mercedes-Benz E430	6.3
	Chevy Camaro SS	5.2
	Ford SVT Mustang Cobra	5.4
SUVs	Ford Explorer	8.0
	GMC Envoy	8.0
	Jeep Liberty	10.0
	Toyota Highlander	8.3

May "the Force" Be with You

For a car to acquire an acceleration, it has to be given a push or a pull. We refer to this push or pull as a *force*, and it is, of course, the engine that supplies it. If you saw the *Star Wars* movies you may think of "force" as something mysterious, but to physicists and engineers it has a precise, technical meaning. An accurate definition of the term was given by the English physicist Isaac Newton more than three hundred years ago. He formulated three laws of motion that now form the basis of almost everything we know about how things move. To understand the first of his laws, let's begin with an object at rest (i.e., something that isn't moving). It will, of course, remain at rest unless we push it, or exert a force on it. This tendency to remain motionless is referred to as inertia. Objects at rest want to stay at rest, which is another way of saying inertia resists motion. Newton incorporated this in his first law, which can be stated as:

A body continues in a state of rest or uniform motion in a straight line, unless a force acts upon it.

Is this believable? Do objects in uniform motion remain in motion indefinitely? If you think about it for a moment, it seems to defy common sense. After all, if you take your foot off the gas pedal, your car slows down and soon stops. Newton's law seems to imply that if you are going 60 mph you will continue at this speed indefinitely without having to apply the gas, and this simply isn't true. It's easy to show, however, that there isn't a problem. The reason your car slows down and stops is that frictional forces and air drag are present. If you could get rid of them, your car would continue with the same velocity forever—and you'd obviously save a lot of gas.

So it takes a force to change the motion of a uniformly moving body, but what acceleration do we get when we apply a particular force? This is answered by Newton's second law:

The acceleration produced by a force acting on a body is directly proportional to the magnitude of the force and inversely proportional to the mass of the object.

Some of the terms in this sentence may be unfamiliar to you. Consider "directly proportional to." This means that if you have a quantity A that is directly proportional to

B, then as A increases B also increases. For example, if you double A, then you double B. Inversely proportional, on the other hand, means as A increases $1/B$ increases; therefore, if you double A, B is cut in half. Another new term here is *mass.* Mass is a measure of the inertia of a body. Roughly speaking, it's the amount of material in a body. Common sense tells us that as a body gets heavier, its inertia increases, and this implies that it gets more massive. It might seem, therefore, that mass and weight are the same, but this is not true. They are, however, related. In fact, weight = mass × acceleration of gravity. And since the acceleration of gravity is roughly the same everywhere on the surface of the Earth (it is 32 ft/sec^2) we can think of weight as a measure of mass. If we were above the surface of the Earth, or on a different planet, the gravity would be different and our weight would be different. Our mass, however, would be the same. In space, for example, an astronaut's mass is constant, but his weight is zero.

Newton's second law can also be written in mathematical form as

$$F = ma,$$

where F is force, m is mass, and a is acceleration. This formula allows us to calculate the acceleration imparted to any object by a force.

Let's take a closer look at this new concept, "force." You exert a force on an object when you push on it. That much seems clear. Surprisingly, though, there is another force involved in this act. Newton showed that there is an equal and opposite force pushing back. He formulated the idea in his third law:

> *For every action there is an equal*
> *and opposite reaction.*

The action he refers to is, of course, the force. The third law therefore tells us that when one body exerts a force on a second body, the second body exerts an equal and opposite force back on it. We see examples of this almost every day. When you hold a garden hose with water pouring out of it, for example, you feel a backward force on your hand. It's also the principle of the rocket. The gas shot out the back of the rocket gives the rocket a forward thrust. It therefore helps Batman (in his rocket-powered Batmobile) rush to the next crime scene.

There appears to be a problem here. If two equal and opposite forces are present, why does the body accelerate? The answer is that the two forces do not act on the *same* body. Suppose you needed to jump-start your Chevy. If you push on it, you exert a force on it; but the Chevy exerts an equal and opposite force on you. The reason the Chevy eventually moves (assuming it does) is that there is another force present, namely, the frictional force between the soles of your shoes and the road. You are, in essence, stuck to the road by this friction and you don't move. The Chevy, on the other hand, is not, and if you push hard enough you eventually overcome the friction associated with the engine, wheels, and so on, and it will move.

Gaining Momentum

Let's assume the force you apply to the Chevy is constant and you push for two seconds. We can easily determine the resulting acceleration. But what if you apply the same force for 3 seconds, or 4 seconds? The acceleration will obviously be greater and the velocity at the end of the time will be greater. This tells us that force, multiplied by the time over which it acts, is important. We call it *impulse,* and designate it by *I.* In mathematical terms,

$$I = Ft.$$

But a given impulse doesn't always produce the same velocity. If you put your entire weight into pushing a Ferrari for 3 seconds, then pushed a heavy minivan with the same force for the same time, the two vehicles would not end up with the same velocity. And it's easy to see why. We know that $F = ma$; if we substitute this into the above formula we get $I = mat$. But velocity is given by $v = at$, and so we find that

$$I = mv.$$

From this we see that the impulse depends on the mass of the vehicle, and the mass of the minivan is greater, so its velocity at the end of the time will be less. The quantity mv in the above formula is given the name *momentum*. This tells us that if an impulse I is applied to a body, it brings about a change in momentum. This makes sense when we think of the force required to stop a car that is in motion. A huge Mack truck is more difficult to stop than a tiny Volkswagen, and if they were in a collision the Volkswagen would definitely come out second best. It's therefore this combination of mass and velocity that is the true measure of the "quantity of motion." I will have more to say about this later.

Managing the Curves

All car fans remember the exciting car chase in the movie *Bullitt*. Steve McQueen, in his Mustang, slid around curves and took breath-stopping jumps on the hilly streets of San Francisco. His straightaway speeds were over 100 miles an hour at times, but how fast did he take the curves? One thing is for sure: McQueen, who did his own stunts, may have felt exhilarated, but

The Isaac Newton School of Driving

he also felt some pretty powerful forces pulling him around inside the car. All car drivers feel a force when they accelerate: they are pushed back in their seats, so even if they can't see out the windows, they know they are accelerating. And when they turn the wheel, they feel another force tending to throw them to the outside of the curve. McQueen no doubt felt this force on every curve. Indeed, if there was anything loose in his Mustang it would have flown across the interior of his car. This is the *centripetal force*.

Newton, in his first law, told us that the car wants to continue in a straight line. When you turn the steering wheel, however, the car changes direction, so unless you're securely attached to the seat by a seat belt, your body will continue moving in the original direction. If you are strapped in, you will feel a force on the seatbelt; this is the centripetal force.

To get an idea of how big this force is, think of a car going around a circular track of radius *R* (fig. 6). Assume that its speed along the track is constant; call it *v*. Its velocity, on the other hand, is constantly changing, since its direction is continually changing. From the diagram

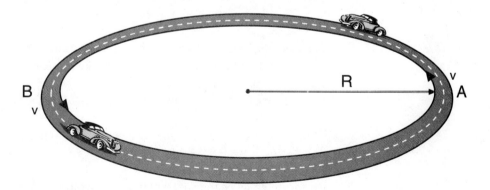

Fig. 6. Automobile on a circular track. Note that its velocity at A is opposite its velocity at B.

we see that at point A it has a velocity v and at point B velocity is v, but it is in the *opposite* direction, so it is $-v$. The change in velocity from A to B is therefore $v - (-v) =$ $2v$. And since acceleration is change in velocity divided by time, we need the time to go from point A to B. The circumference of the circle is $2\pi R$, so the distance from A to B is half of this, or πR. But velocity = distance/time, so time = distance/velocity. This means that it takes a time t to travel a distance πR at speed v, where $t = \pi R/v$. Using this in our expression for acceleration, we get

$$a = \text{(change in velocity)/time} = 2v/t$$
$$= 2v/(\pi R/v) = (2/\pi)\, v^2/R.$$

It's important to note, however, that this is the *average* acceleration. We prefer instantaneous acceleration. Leaving out the details (they are a little complicated), we get

$$a = v^2/R.$$

The centripetal force associated with this acceleration is

$$F = mv^2/R.$$

Let's use this formula to calculate the acceleration for several circles of various radii. We would like the velocity in miles per hour and the acceleration in ft/sec^2 so we will have to use our conversion factor (15/22). Our formula for velocity is then

$$v = [15/22\ a(\text{ft/sec}^2)R(\text{ft})]^{\frac{1}{2}}.$$

This formula gives the speed in miles per hour for various accelerations around a curve of radius R. It also, of course, applies to any turn, which is a section of a curve.

It is most convenient to present the results in the form of a table (see table 2). Note that in a few cases I have indicated the acceleration in g's. As we saw earlier, one g is about the maximum your tires could withstand.

You can use table 2 in several ways. For example, say you were rounding a curve of radius 100 feet at 20 mph. The force on your body would be approximately ¼ g, or one-quarter of your body weight. If you weighed 180 pounds, it would be 45 pounds. The table also represents the maximum speed that can be achieved for turns of various radii. For example, if the curve you are on has a radius of curvature of 150 feet, the maximum speed you should go if you do not want to exceed ½ g is 35.61 mph. McQueen may have got to ½ g, but I'm sure he never exceeded 1 g. He wouldn't have been able to finish the film if he did.

Torque

"Torque" is a word you hear frequently in relation to high-performance cars. As the ads say, "The torque of the Porsche 911 Turbo is 413 lb-ft at 2700–4600

Table 2. Speed and acceleration for curves of various radii

	Speed (miles/hour)				
a (ft/sec²)	R = 50 ft	R = 100 ft	R = 150 ft	R = 200 ft	R = 300 ft
4	10.92	14.54	17.80	20.56	25.18
8 (¼ g)	14.54	20.56	25.18	29.08	35.61
12	17.80	25.18	30.84	35.61	43.62
16 (½ g)	20.56	29.08	35.61	41.12	50.36
24	25.18	35.61	43.62	50.36	61.68
32 (1 g)	29.08	41.12	50.36	58.16	71.23

Note: a = acceleration; R = radius.

rpm. The torque of the Ferrari 360 Modena is 275 lb-ft at 4750 rpm."

We saw earlier that a force gives a body straight-line motion. A torque, on the other hand, gives a body rotational motion. A torque wrench, for example, gets a screw to move around in a circle, and when you are trying to get the top off a jar of fruit, you are applying a torque.

The magnitude of a torque depends not only on the force that's applied, but also on the distance between the center of mass of the body and where it is applied. The center of mass, or center of gravity, as it is sometimes called, is the point within the body at which all the mass could be considered to be concentrated.

With this knowledge we can define torque mathematically. Letting τ be torque we can write

$$\tau = Fd.$$

F is the force applied and d is the perpendicular distance from the center of mass to the line of force as shown in the diagram (fig. 7). More generally, d can also be the distance to the fixed point of the body. From the formula we see that the larger the distance from the applied force, the larger the torque. This is why we place the doorknob as far as possible from the hinges of the door; in doing this we are maximizing the torque on the door and making it easier to open.

In practice, part of the body doesn't need to be pinned down for a torque to be applied. In the collision between two cars, for example, if one of the cars hits the other near the front it will impart both translational (straight line) motion and rotation to the second car. This means that both torques and forces are applied. If the collision occurs when someone runs a red light at high speed, the torque at impact can be extremely high.

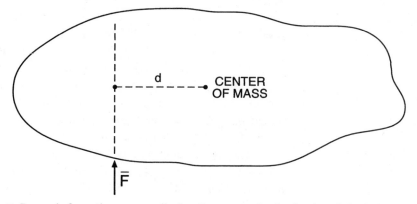

Fig. 7. Torque is force times perpendicular distance to the fixed point of the body. Here we assume the fixed point is the center of mass.

Energy

Imagine that you're on your way to work in your new Corvette. After parking it, you take the elevator to your office. You're on the phone for the first hour, then you read through some contracts and sign them. You spend some time thinking about how to land a particularly lucrative new contract. By noon you're exhausted and you compliment yourself on how hard you've worked. It would probably disappoint you to hear that according to a physicist's definition, you haven't done a bit of work.

Work is, indeed, an important concept in physics. It is defined as work = force × distance, or in mathematical terms,

$$W = Fd,$$

where F is force and d is the distance over which the force is applied. Note that this is different from the definition of torque in that the distance is not a perpendicular distance. Looking closely at the formula we see that physics seems to clash with common sense here. For example, it implies that if you lift a box of tools out of the trunk of your car, you do work against the force of gravity; but if you carry the box the same distance horizontally, you do no work at all. On the other hand, if you pushed the box across the floor, you would do work against friction. This is not our usual concept of work, so you have to be careful in dealing with it.

Closely related to work is *power,* and when we talk about cars we frequently talk about power, or more exactly, horsepower. The horsepower of the Cadillac DeVille is 398 at 6400 rpm. The horsepower of the Mercedes-Benz C320 is 215 at 5700 rpm. What exactly do these numbers mean? Let's begin with power; it's defined as the rate of doing work, in other words, it's how much work gets done in a particular time. In mathematical symbols it is $P = W/t$, where P is power and W is work. From this we can write $P = Fd/t = Fv$. The units of work are force (pounds) × distance (feet), or pound-feet (lb-ft), so the unit of power is pound-feet per second.

Where does the "horse" part come in? It goes back to 1783 when the Scottish engineer James Watt decided to see how much power a horse could generate. After experimenting, he determined a strong horse could raise a 150-pound weight approximately 4 feet in a second. He therefore defined the horsepower to be 550 pound-feet per second.

You have likely seen another unit of power on your electric bills. It is the watt (or kilowatt, which is 1000 watts); named for James Watt, it is a metric unit. One horsepower is 746 watts. It's easy to show that if you kept a horse in your backyard, you could use it to light up approximately twelve lightbulbs in your house.

Work is closely related to another important concept

in physics called *energy*. If you do a certain amount of work on an object, you can set it in motion. An object in motion has energy, which is usually called *kinetic energy*. We can determine it as follows:

$$W = Fd = mad = ma(\tfrac{1}{2}at^2) = \tfrac{1}{2}m(at)^2 = \tfrac{1}{2}mv^2.$$

The kinetic energy of an object of mass m moving at speed v is therefore $\tfrac{1}{2}mv^2$. A car in motion has kinetic energy, and if you know its mass and velocity you can calculate it. Its units are the same as those of work.

Conservation of Energy

Kinetic energy is just one of many forms of energy. To see another form, consider a ball that is thrown up in the air. When it is first thrown, it has a certain amount of kinetic energy, but as it rises against the force of gravity, its velocity slows and its kinetic energy decreases. Finally, when all the kinetic energy is depleted, it stops. Is the energy lost? No, the ball now has a different kind of energy—energy of position. In effect, its kinetic energy of motion has been transformed into energy of position. We refer to this energy as *potential energy*. As the ball begins to fall back to Earth it gains kinetic energy. Finally, when (or just slightly before) it strikes the Earth it has only kinetic energy; all its potential energy is used up. We define potential energy by

$$P = mgh,$$

where h is height above the Earth.

This means that one type of energy can be transformed into another type, and that no energy is lost in the process. When a car flies off a cliff it has mostly potential energy (assuming its speed wasn't too high).

Just before it strikes the ground, however, all this potential energy is transformed into kinetic energy. This seems to imply that energy in general cannot be destroyed, and indeed this is the essence of an important principle in physics called the *principle of conservation of energy.* This principle tells us that energy cannot be destroyed or created; it can only be changed in form.

In the above example we saw the transformation of energy from kinetic to potential, but what happened to the energy when the car struck the ground? It appears at first that it has disappeared. As I mentioned earlier, however, there are many forms of energy. If you look closely at the ground where the car struck, you will see that it is depressed. Some of the energy has gone into depressing the ground and in smashing up the car. This is deformational energy. Furthermore, if you measured the temperature of the soil where the car hit, you would find that it is slightly higher than it was earlier. So some of the energy has been transformed into heat energy.

There are, in fact, several other forms of energy. Electrical energy is one you are no doubt familiar with. And in relation to cars an important one is chemical energy. It is the energy given off in a chemical reaction, such as the burning of gasoline. Another example is sound energy. Electrical energy, for example, can be converted to sound energy using a microphone. Light is another form of energy.

The Power of Physics

Physics can, indeed, tell us a lot about how cars react to forces. It also shows us that power, energy, and momentum are important in relation to cars. As a slightly more complicated example, let's consider how the distribution of weight in a car changes as the car accelerates and decelerates. We know that the weight of a car is approximately evenly distributed when the car is not moving. In other words, half of the weight is on the front axle and half on the back axle: for a car of 3000 pounds, 1500 would be on the front axle and 1500 on the back, though it actually varies slightly, depending on the model of the car.

The shifting of weight when a car accelerates or decelerates is important to the race driver. Good race car drivers always make a quick estimate in their head of how the weight of the car is being shifted. Without this knowledge they could easily oversteer or understeer.

Since we're not as good at making rough estimates in our head as most good race drivers, we'll calculate it. The calculation may involve a little more mathematics than you're used to, but I'll try to keep it as simple as possible. We'll start with the forces on a car that is braking (fig. 8). As shown in the figure, several forces act on a car. Its weight W acts downward from its center of mass. It is counteracted by two forces acting upward on the tires. We'll call them F_1 and F_2 and assume they are equal. If the car is standing still, F_1 and F_2 will be equal to W, the weight of the car. The two forces f_1 and f_2 are frictional forces on the tires.

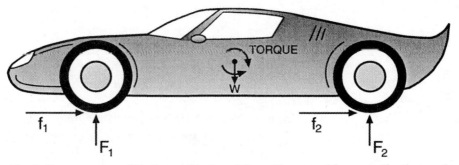

Fig. 8. Forces on a car: W is its weight, F_1 and F_2 are the upward forces on the tires, and f_1 and f_2 are the frictional forces.

Now, what happens if the car is moving and we apply the brakes? This will cause F_1 to increase and F_2 to decrease; in other words, the front end of the car will get "heavier" and the back end will get "lighter." The reason is that the frictional forces f_1 and f_2 act to slow the car at ground level, but the inertia of the car, which tends to keep it moving, acts at the center of gravity, and it is several inches above the ground. The result is a torque that tries to topple the car on its nose.

If the car doesn't topple, this torque—which is in the anticlockwise direction in the figure—is balanced by a torque in the clockwise direction. We can calculate these torques if we know the weight of the car W, the wheelbase R, and the height of the center of mass h. Equating the two torques and using $F_1 + F_2 = W$ gives

$$W_d = Fh/R.$$

We have changed notation slightly here; W_d is the weight that is shifted, and F is the force on the vehicle, given by $F = ma$.

As an example, assume our car weighs 3000 pounds, $R = 100$ inches, and $h = 24$ inches, and the acceleration is 32 ft/sec^2. We get

$$W_d = (3000/32) \times 32 \times (24/100) = 720 \text{ pounds.}$$

This tells us that 720 pounds are transferred. Since we originally had 1500 pounds on each axle, we now have 2220 pounds on the front axle and 780 pounds on the back.

With this much weight added to the front end, we would have trouble steering. We would, in fact, have a tendency to oversteer if we tried to turn the car. Race drivers know all about this when they apply the brakes, but most people don't. As we also saw, when we accelerate there is a weight transfer to the rear of the car, in which case we would have a tendency to understeer. A similar weight shift occurs when we round a corner. Cornering in one direction shifts the weight in the opposite direction.

It is easy to see from the calculations that weight transfer is less if the center of gravity is lower. This is why cars with low centers of gravity generally handle better under conditions of braking and steering. Control of weight transfer is particularly important in avoiding sliding (exceeding the adhesive limits of the tires) and flipping the car over, or onto its side.

This information shows us that physics is particularly useful in making predictions about the motion of a car. It is, in fact, indispensable.

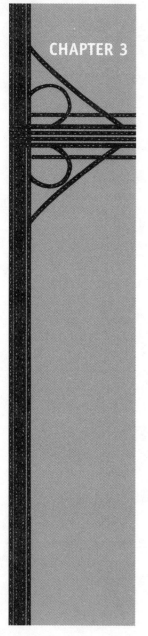

CHAPTER 3

All Revved Up

The Internal Combustion Engine

"Vrrrooooooomm! Vrrrooooooomm!" You rev up your engine several times just to hear its power. It's a great sound, and a great feeling—knowing that you have so much power. But it wasn't always so. The earliest cars had engines that had only one or two horsepower.

The first working internal combustion engine, a noncompression engine that used illuminating gas as fuel, was built by Jean Joseph Lenoir of France in 1860. It had a low efficiency (about 5%) and required extensive cooling, so only a small number were sold. But it is important for another reason. In 1863 Lenoir hooked one of his engines to a small "wagon" and made a short trip around Paris. It could hardly be called an automobile—they wouldn't come for another twenty years—but it was the first "horseless carriage."

All early engines were two-stroke noncompression engines; in other words, there were two piston movements over one revolution of the engine for each cycle. The first four-stroke compression engine, which is the main type used today, was the brainchild of Nikolaus Otto of Germany. His engine had four strokes or four piston movements over two engine revolutions for each cycle.

Otto started out as a salesman, but one day he came across an article on Lenoir's engine in the newspaper, and after reading it he became intrigued with the engine's possibilities. Soon he was spending most of his spare time working on engines. He realized almost immediately that there were flaws in Lenoir's engine and he was determined to improve on it. He experimented with compressing the fuel, but his first experiments scared him so much that he left the idea alone for several years.

Otto was enthusiastic but he had few funds. In 1864, however, fate smiled on him. Eugen Langen, a successful businessman, came to see his engine. Langen was fascinated by it and within a short time he had raised enough money to form a company. On March 31, 1864, the first manufacturing company for internal combustion engines came into being. It was three years, however, before they had ironed out all the problems.

In 1867 they took their pride and joy to the Paris Exposition. A gold medal was to be awarded to the best engine. Most of the judges were not impressed with Otto's entry, but when they began looking at the efficiency of the various engines it soon became obvious that Otto's engine was superior to the others. It used only half as much fuel and produced more power, and so the judges awarded it the gold medal.

Langen and Otto moved their factory to Deutz, a suburb of Cologne, in 1872. Langen hired Gottlieb Daimler, an engineer with excellent credentials, as production manager. Over the next few years they worked diligently to improve the engine. But there seemed to be a problem. No matter what they did with the two-cycle model, they couldn't get much more than three horsepower out of it.

Otto remembered his early experiments with the compression engine and soon decided that the fuel-air mixture had to be compressed. He also began experimenting with a four-cycle engine with an intake cycle, compression, ignition, and an expansion and exhaust cycle over four piston strokes rather than two. Of particular importance was that it all took place in the same cylinder over two crankshaft revolutions. Because this innovation went against the trend at the time, both Langen and Daimler thought Otto was wasting his time and they were sure nothing would come of his "crazy" idea. But when Otto demonstrated a prototype model to them in late 1872, they were impressed.

It was soon obvious that the new four-cycle engine had many advantages over the old two-cycle model. The 3-horsepower limit was quickly overcome, and over the next few years tremendous advances were made and sales of the engine increased dramatically.

Although there was considerable discussion about using the new engine in a horseless carriage, that vehicle would not come for another few years. Daimler, however, was intrigued with the idea. After a disagreement in 1882 he left the Deutz plant and formed his own company. He took with him one of the best engineers, Wilhelm Maybach.

Gasoline was now being used as fuel, and Daimler was convinced it would be the ideal choice for the new horseless carriage that he visualized. He began working on the design of a chassis while Maybach worked to perfect the engine. By 1886 they had their first automobile. In September of that year, Daimler took it for a test drive around his property. It had a water-cooled engine of 1.1 horsepower and rotated at 650 rpm.

At about the same time, Karl Benz was working on a similar model. His first, a three-wheeled model, was presented to the public in 1888, but it was not a success. In 1893, however, Benz introduced a four-wheeled model with a much larger horsepower engine, and by 1899 he had sold two thousand of them. Daimler and Benz are now considered to have been the first to produce and sell automobiles.

It didn't take long for the idea to catch on in America. The first to get into the act were Charles and Fred Duryea. They built their first "Duryeas" in 1892, and by 1896 they had produced and sold thirteen. America, however, was not ready for the noisy and somewhat unreliable combustion engine. Though it was shown in the first national auto show in New York in 1900, it was the quieter and more reliable electric car that attracted more attention. The loud backfiring and clanging of the combustion engine scared most people. By the third show in 1903, however, significant improvements had been made, and the combustion engine was center stage.

By this time, the man who was to dominate the auto scene for the next few decades was quietly working on his first model. In 1896 Henry Ford produced his quadracycle, so called because of its narrow, bicycle-like tires.

Soon afterward, Ford formed an automobile company, but it didn't last long. A stroke of luck, however, soon put him back in business. One of the cars he had produced was a racing car, and in October 1901 a race was held just outside Detroit. Alexander Winton, a Cleveland manufacturer of cars, was heavily favored to win, and Ford entered the race as an underdog. Winton's car had difficulties during the race, however, and Ford won. As a result, he soon had several backers for a new company.

Within a few years he had produced his famous Model T and the rest is history. By 1908 he had sold 10,000 Model T's.

What Makes It Run? The Four-Cycle Combustion Engine

Before we get into the details of how the four-cycle engine works, let's look at the major components of the engine. In figure 9 we see the head, the block, and the crankcase. The head contains the mechanisms for opening and closing the valves that let in the fuel-air mixture and allow the exhaust gases to escape. The central part of the engine is the block that contains the cylinders, inside of which are the pistons. At the top of the cylinder is the spark plug that is used to ignite the fuel-air mixture. The piston fits snugly inside the cylinder where it moves easily up and down (fig. 10). On the outside of the piston are rings that help seal the area inside. A connecting rod connects the piston to the crankshaft. Two other important openings in the cylinder are those for the intake and exhaust valves. And finally, beneath the block is the crankcase. It houses the crankshaft and the oil pan.

In almost all engine specifications, you will see some mention of bore and stroke. Both are important operat-

Fig. 9. Cross section of engine showing head, block, and crankcase.

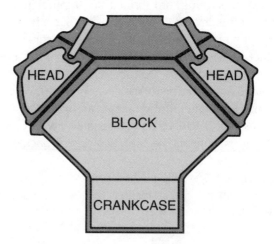

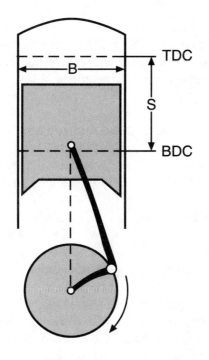

Fig. 10. A piston moving up and down in a cylinder. TDC is top, dead center (referred to as T_c). BDC is bottom, dead center (referred to as B_c).

ing parameters that determine the power of the engine. Bore is the inside diameter of the cylinder, and stroke is the distance the piston moves from top to bottom. Both are shown in figure 10. The following are a few representative values, with bore sizes given in inches and millimeters (mm), and strokes given in inches and millimeters (all the vehicles are from 2002):

Model	Bore (in/mm)	Stroke (in/mm)
Chevy Corvette Z06	3.90/99	3.62/92
Audi A6	3.32/84	3.66/93
BMW 540i	3.62/92	3.26/83
Lexus GS 430	3.58/91	3.25/83
Ford Focus ZTS	3.39/85	3.52/88
Jaguar ZK8	3.39/86	3.39/86
Mercedes-Benz CLK 430	3.54/90	3.31/84

The cylinders can be lined up in several different ways. In figure 9 they form a V, the most popular of this type being the V8 engine with eight cylinders. Another common arrangement is the in-line or straight, where the cylinders are all in a line. Other arrangements that have been used are "opposed cylinders" with the cylinders opposite one another, and cylinders in a W arrangement and mounted radially. The radial mounting is used in aircraft engines. We will talk mostly about the straight and the V8 since they are the most common ones used in cars.

Getting back to the four-stroke engine, we see from figure 10 that the piston moves from a position where the volume above it is minimum to one where the volume is maximum. The upper position is referred to as top, dead center, and the bottom as bottom, dead center. We will refer to these as T_c and B_c.

In the first, or intake, stroke the piston travels from T_c to B_c with the exhaust valve closed. Since air can't get past the rings, a vacuum is created in the region above the cylinder. When the intake valve is opened the fuel-air mixture rushes in to fill the vacuum. When the piston reaches B_c the intake valve closes and the piston moves back up, compressing the fuel-air mixture in the small space between the top of the piston and the top of the cylinder. The compression raises both the pressure and temperature of the mixture. Near the end of the stroke, the spark plug is fired and the fuel-air mixture is ignited. The resulting explosion creates considerable pressure that forces the piston back down, creating what is called the power stroke. During this stroke, power is transmitted from the piston to the crankshaft, which in turn transfers it through the transmission, driveshaft, and so on to the wheels. (See fig. 11.)

Near the end of the power stroke the exhaust valve is opened and most of the exhaust exits. There is still

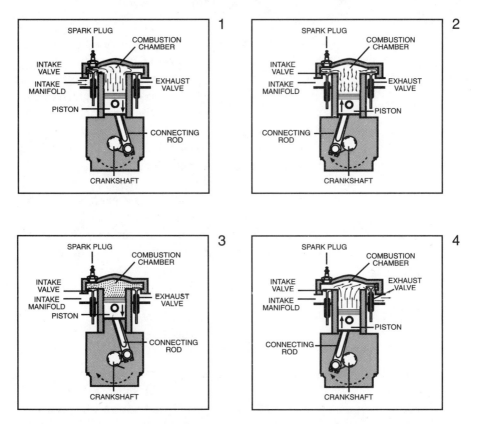

Fig. 11. Four-stroke combustion engine: intake stroke (*upper left*); compression stroke (*upper right*); power stroke (*lower left*); exhaust stroke (*lower right*).

some exhaust left at atmospheric pressure, however. It is blown out during the last of the four strokes, the exhaust stroke. In this case the piston moves back up to T_c and the cycle starts over again.

There are, of course, several cylinders and pistons in any car (usually six or eight) and they do not all fire at the same time. While one cylinder is going through stroke one, another will be going through stroke two, and so on. The exact sequence is referred to as the *firing order*.

Although we won't be dealing with it much here, I will briefly discuss the two-cycle engine. It is used in small engines such as those in lawn mowers, chainsaws, and small boats. The major differences from a four-cycle engine are that it has no moving valves, and the spark plug fires each time the piston hits the top of the cylinder. When the piston is at the bottom, an intake port is uncovered, allowing the fuel-air mixture to flow from the crankcase to the cylinder. It has been partially pressurized in the crankcase. Further compression occurs as the piston rises. When the piston reaches the top, the fuel-air mixture is ignited by the spark plug. At the same time as compression occurs in the cylinder, a vacuum is created in the crankcase and fresh fuel is drawn in. Ignition of the fuel creates the power stroke, and the piston is forced downward. The cycle then repeats itself.

But Is It Efficient?

When you pick up an automobile magazine at one of the newsstands, one of the first things you notice in it are page after page of specifications on new models of cars. We already discussed bore and stroke, but these lists also contain such things as displacement, compression ratios, horsepower, and torque. Let's look at them.

We'll begin with the speed of the piston. It moves back and forth in the cylinder with a speed of about 15 ft/sec to about 50 ft/sec. Why is it limited to 50 ft/sec? There's a good reason. At this speed there is considerable strain on both the piston and the connecting rod. If you tried to force it to higher speeds, something would likely give, and you'd be in trouble. If you've ever "thrown a rod" you know what I mean. Many years ago, some of us were traveling to a track meet in an older car owned by one of the track team members. I didn't like the sound of the engine almost from the beginning but said nothing to the driver. I began to get worried, however, when the noise became much louder late that night. We were miles from the nearest town and there was practically no traffic on the road. Suddenly I heard a loud crash and clanking from the direction of the engine. I'm sure I lifted my feet, worried that something was going to come flying through the floorboard. It was soon obvious that we had thrown a rod, and since we were in the middle of nowhere, we ended up sleeping in the car for the night.

Another reason that the speed of the piston is limited is that the fuel-air mixture brought in through the intake valve can move only so fast. To be increased, the size of the intake valve would have to be increased, and these valves are already at maximum size.

With the above piston speeds, the crankshaft makes from 500 to 5000 revolutions per minute (rpm), with typical cruising speeds giving about 2000 rpm. Large engines usually operate in the range of a few hundred rpm, while small ones such as those in model airplanes can have speeds of 10,000 rpm or more.

As the piston moves between T_c and B_c, a certain volume is displaced. It is referred to as the *displacement* of the engine. The total displacement, which is the number usually given for engines, is the sum of the volumes of all cylinders. Displacement is a good indication of the engine size. It is usually given in liters (l), cubic centimeters (cc), or cubic inches (ci), where 1 l = 1000 cc = 61 ci. Typical displacements for modern cars are 2 to 5 liters (2000 to 5000 cc). Trucks frequently have displacements of well over 5 liters. The displacements of several engines are given in the following table.

Model	Displacement (ci/cc)
Chevy Camaro SS	346.9/5665
Ford SVT Mustang Cobra	280.8/4601
Audi A6	254.6/4172
BMW 540i	267.9/4391
Lexus GS 430	261.9/4293
Dodge Stratus ES	166.9/2736
Honda Accord EX	183/3000
Chevy 2500 HD 2WD (truck)	495.8/8127

Also important in relation to cars is the *compression ratio*. It is a measure of the amount of pressure applied to the gas mixture in the combustion chamber. Numerically, it is the volume of the combustion chamber at T_c compared to its volume at B_c. Over the years, compression ratios have increased considerably. Early cars (1920 to 1940) had compression ratios in the range of 4 to 5 while values as high as 10 are common now. The modern range is approximately 8 to 11. The compression ratios of several cars are given in the table below.

Model	Compression Ratio
Chevy Corvette Z06	10.5:1
Audi A6	11.0:1
BMW 540i	10.0:1
Lexus GS 430	10.5:1
Mercedes-Benz E 430	10.0:1
Chevy Camaro SS	10.1:1
Ford SVT Mustang Cobra	9.9:1

If you've ever driven at high altitudes, you know the air pressure affects the output power of an engine. Reduced air pressure affects people—they get dizzy and lightheaded at high altitudes—and it only stands to reason that it would affect cars. I had a sabbatical on the Big Island of Hawaii several years ago, and visited the observatory at the top of Mauna Kea many times. It has an altitude of 13,796 feet. (Actually, from its base beneath the ocean it is 33,476 feet high—higher than Mount Everest.) The trip to the top in a car was always an interesting experience. Long before you were anywhere near the top you were in low gear, and at times it seemed like you might be able to walk faster than the car was moving. Modern fuel ignition systems do, indeed, measure the air and fuel density, but they can't restore the horsepower that's lost because of the low density of oxygen molecules in the cylinder. So whenever you go to a high altitude, you should expect to see some power loss in your car.

The major function of an engine, of course, is work. As we will see in a later section, one of the easiest ways of determining work is to make a plot of pressure versus volume. To see why, let's look at the definition of work:

$$\text{work} = \text{force} \times \text{distance}.$$

We can rewrite the right hand side as [(force)/(area)] × volume. But this is just pressure × volume, or *PV*. *PV* diagrams are also called indicator diagrams. They can easily be obtained by hooking an oscilloscope up to an engine.

In practice, the work delivered to the crankshaft is usually considerably less than that given by *PV*, so we distinguish *PV* by calling it *indicated work* W_i, and call the actual work delivered to the crankcase *brake work* W_b. The ratio of these is referred to as the *mechanical efficiency* E_m of the engine:

$$E_m = W_b/W_i.$$

Mechanical efficiencies of modern cars are typically in the range of 75% to 95% at a wide-open throttle. They decrease with decreasing engine speed.

A particularly good way of comparing engines is through what is called *mean effective pressure* (mep). The pressure in the cylinder is continually changing throughout the cylinder, but we can take an average. It is useful because it is independent of the size of the engine. It is defined as

$$mep = (\text{work in one cycle})/(\text{displacement volume}) = W/V_d.$$

Again, because of frictional losses and so on we have indicated mep and brake mep.

You are no doubt quite familiar with the terms *torque* and *power*. They are the two major terms used when discussing the engine's merits. In the last chapter we saw that torque is force multiplied by some lever arm. It's a good measure of the engine's ability to do work since it gives the amount of twist or turning power of the engine. The units used are N-m or lb-ft where 1 N-m = .738 lb-ft. If the number is given without units, it is usually assumed to be lb-ft. Torque varies with the engine speed so the rpms have to be specified when giving it (fig. 12). Typically, torques are in the range of 150 to 400 lb-ft with engine speeds of 3500 to 6000 rpm, with a few exceptions. Table 3 gives the torques for several different types of vehicles.

Because torque varies with engine speed, most manufacturers try to flatten the curve as much as possible, giving a more uniform torque over the range of speeds. The point of maximum torque is referred to as the *maximum brake torque speed.*

Power is the rate at which an engine does work. Like torque it is also a function of engine speed, so the rpms

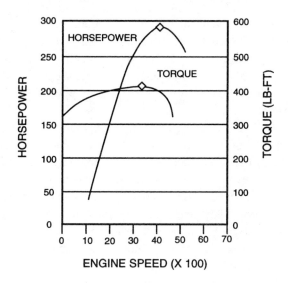

Fig. 12. Horsepower and torque versus engine speed.

Table 3. Torque for several 2002 vehicles

Vehicle Type	Model	Torque (lb-ft)
Sports and luxury cars	Chrysler Crossfire	270 @ 4000 rpm
	BMW Z3	214 @ 3500 rpm
	Ford Thunderbird	267 @ 4300 rpm
	Lamborghini Murciélago	479 @ 5400 rpm
	Cadillac Escalade EXT	380 @ 4000 rpm
	Chevy Corvette Z06	385 @ 4800 rpm
Family sedans	Ford Focus ZTS	135 @ 4500 rpm
	Dodge Stratus	192 @ 4300 rpm
	Honda Accord EX	195 @ 4700 rpm
	Hyundai XG 300	178 @ 4000 rpm
SUVs	Jeep Liberty	235 @ 4000 rpm
	GMC Envoy	275 @ 3500 rpm
	Ford Explorer	255 @ 4000 rpm
Trucks	Chevy Silverado	520 @ 1800 rpm
	Chevy Tahoe Z75	325 @ 4000 rpm

have to be specified. The most common unit of power is horsepower (hp), but kilowatts (kw) are also used. The range of powers in automobiles is from 50 to 350 hp (except for high-performance sports cars); trucks and large SUVs usually have higher powers. A few examples are given in table 4.

As if the Dodge Viper needs more than the 500 hp it already has, Houston-based supermechanic John Hennessey has been increasing the horsepower of some Vipers beyond the 800 hp mark. He has actually taken one model up to 830 hp. When you consider that it takes all of about 50 hp to power the vehicle at 60 mph on the level, 850 hp is a lot of power by any standards. Hennessey does this using two industrial-sized turbochargers. We'll talk about turbocharging a little later.

Not to be outdone, the Bugatti EB 16, a sixteen-cylinder turbocharged engine that is supposed to make its appearance in 2003, is scheduled to have a horsepower

Table 4. Horsepower for several 2002 vehicles

Vehicle Type	Model	Horsepower
Sports and luxury cars	Audi A6 4.2	300 @ 6200 rpm
	BMW 540i	282 @ 5400 rpm
	Chevy Camaro SS	325 @ 5200 rpm
	Ford SVT Mustang Cobra	320 @ 6000 rpm
	Jaguar × type	231 @ 6800 rpm
	Lamborghini Murciélago	571 @ 7500 rpm
	Dodge Viper RT 10[*]	500 @ 5900 rpm
	Chrysler Sebring	200 @ 5900 rpm
Family sedans	Nissan Sentra	122 @ 6000 rpm
	Chrysler PT Cruiser	155 @ 5500 rpm
	Toyota Camry Solara	198 @ 5200 rpm
SUVs	Jeep Liberty	210 @ 5200 rpm

[*]A 2003 model.

of 987. It will presumably go from 0 to 60 in 3 seconds! Where will it all end?

So much for horsepower. Let's turn now to the efficiency of the engine. All drivers, and particularly race drivers, are concerned about the efficiency of their engines, or if they aren't, they should be. As it turns out, there are several different efficiencies that we need to be concerned with. Let's begin with *combustion efficiency*. You might think that every drop of gasoline that goes into the combustion chamber is burned, but it isn't. A small fraction actually exits with the exhaust. If the engine is operating properly this amounts to about 2% to 5%. In this case we say the combustion efficiency E_c is .95 to .98.

Next, there is *thermal efficiency* E_t. Even if all the fuel is burned, not all the fuel's energy is converted to rotational energy. The chemical energy in a pound of gasoline is 19,000 to 20,000 BTUs. Of this, only about one third usually ends up as usable horsepower. This is referred to as the *thermal efficiency,* and it is caused by such variables as combustion chamber design, compression ratio, and timing. A regular car might have an E_t of .25. Race car engines, on the other hand, have E_t's typically in the range of .35. Again, we have both indicated and brake E_t. Indicated E_t might be in the range of .5 to .6.

Earlier we saw that mechanical efficiency can be defined in terms of work. It turns out it can also be defined in terms of indicated and brake thermal efficiencies:

$$E_m = (E_t)_b \: / \: (E_t)_i.$$

Physically, mechanical efficiency is a measure of the power it takes to overcome friction in the engine and to run the engine accessories such as the water and oil pumps. If you measure the power of an engine using a dynamometer, the ratio between it and the workable power in the cylinder is the mechanical efficiency.

Of particular interest to engine enthusiasts is what is called *volumetric efficiency* E_v. Suppose you have an engine with a cylinder that has a volume of 120 cubic inches when the piston is at the bottom of the cylinder (at B_c). You assume that when the piston is at this position during its cycle that it will have taken in 120 cubic inches of fuel-air mixture. But this isn't the case. Because of several things such as the vacuum in the manifold, the amount of the fuel-air mixture that fills the space will be less than 120 cubic inches. The ratio of the amount of fuel-air that actually fills the chamber to the amount of air at atmospheric pressure that would fill it is the volumetric efficiency. If 120 cubic inches of fuel-air mixture were in it, it would have 100% efficiency. Most engines have E_v's of 80% to 100%. We usually specify E_v at wide-open throttle; it decreases for partially open throttle. It's important to note that volumetric efficiency is also a function of engine speed (fig. 13).

Superchargers and Turbochargers

We saw that the Dodge Viper can be increased in horsepower from 500 to over 800 using turbocharging, and

Fig. 13. Volumetric efficiency versus engine speed.

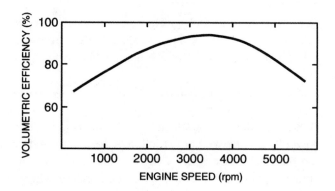

The Isaac Newton School of Driving

the Bugatti, which will have a horsepower of 987, is also turbocharged. What is turbocharging? The idea behind both turbocharging and supercharging is improving the volumetric efficiency. As we saw in the last section, volumetric efficiency can be improved if we get more fuel-air mixture into the chamber. One way of doing this is to pressurize it by using a compressor mounted on the intake system. More air and fuel enter the cylinder during each cycle when this is done and E_v increases; in fact, the overall power of the engine increases.

Superchargers are compressors that are driven directly by the engine—usually by way of a pulley off the crankshaft. This is undesirable because it puts a load on the engine output. Superchargers have an advantage over turbochargers, however, in that they have a quick response to changes in the throttle. One of their disadvantages is that in increasing the pressure of the air that is coming in they also increase its temperature. This is undesirable because it can cause pre-ignition and knocking. To avoid this problem, most superchargers are equipped with aftercoolers that cool the compressed air.

Turbochargers get around one of the main problems of superchargers. They are not driven by the engine and therefore do not put a load on it. In this case compression is a result of a turbine mounted in the exhaust system of the engine. The hot exhaust gases leave the exhaust manifold in their usual manner, but instead of exiting they are passed through a turbine or fan that is set in motion by the passing gases. This fan is connected to a compressor.

Although turbochargers don't work off the engine, they do cause a more restricted flow of the exhaust gases, which in turn causes a slightly higher pressure in the cylinder exhaust port. A slight decrease in power results.

A disadvantage of turbochargers, compared to superchargers, is what is called the turbo lag—a lag that occurs with sudden throttle changes. Like superchargers, turbochargers must also be equipped with aftercoolers.

The Porsche is renowned for its turbo models. The Porsche 911 Turbo has a top speed of 189 mph, and the newer Porsche 911 GT2 can reach 196 mph. The acceleration of the GT2 is 0 to 60 in 4.1 seconds. Toyota's Camry Solara is an example of a car that is supercharged. The Rousch Stage 3 Mustang and the Chrysler Crossfire are also supercharged. Turbocharging is particularly effective in diesels and can increase the power by 50% while lowering fuel consumption. (See fig. 14.)

It's Hot: Heat Transfer in Engines

Internal combustion engines obviously get very hot, and some of this heat must be dissipated as soon as possible after it is generated. Of the overall energy that

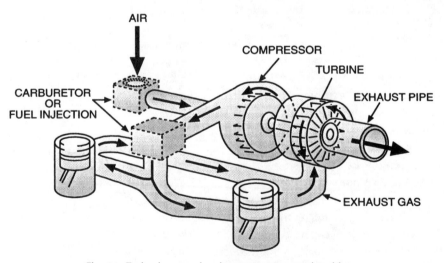

Fig. 14. Turbocharger, showing compressor and turbine.

enters the engine, about 35% is converted to useful crankshaft work and about 30% is lost in the exhaust. This leaves 35% that we must get rid of through heat transfer.

Temperatures within the combustion engine can reach several thousand degrees (fig. 15). If the materials and oil in this region were exposed to temperatures of this magnitude for long they would soon break down. Removing heat from this region is therefore critical. It is, of course, just as important to remove only the required amount, as the engine should operate as hot as possible for maximum efficiency.

The hottest regions of the combustion chamber are near the spark plug and exhaust valves. Temperatures are typically 600° C or higher here. The face of the piston is also a hot region. Unfortunately, these are all difficult regions to cool.

Three modes of heat transfer occur within this region: conduction, convection, and radiation. Conduction is one of the most important. If you have a metal rod and

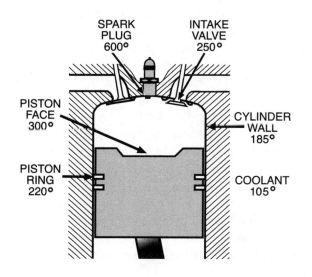

Fig. 15. Temperatures (in degrees Centigrade) at various points inside the cylinder and along the piston.

place one end of it in a flame you know that the other end eventually gets hot, even though it isn't in contact with the flame. The reason is that the flame transfers energy to the atoms it is in contact with, and they soon begin to vibrate more rapidly. Their vibrational energy is then transferred to neighboring atoms. In this way heat, or heat energy, is passed from atom to atom.

A measure of thermal conductivity is called the *coefficient of conductivity*. Metals such as silver and copper have high conductivities. Materials such as concrete, ceramics, and air have very low conductivities and are referred to as *heat insulators*.

A second mode of heat transfer that is also important in engines is convection. In this case, either gas or liquid is moved bodily from one place to another. For example, a hot air furnace uses convection: air is heated, then transferred to various rooms in a house. Heated material that is forced to move by a blower is forced convection, and if it moves naturally on its own it is free convection. Rising hot air is a good example of free convection.

The last of the three methods, radiation, is not as important in engines as the other methods, but it does occur. If an object is heated red-hot you can easily feel the heat with your hands from some distance away due to radiation. Any heated object gives off radiant energy in the form of electromagnetic waves, which travel at the speed of light and do not need a medium to propagate them. They are transmitted in a vacuum as well as air. Our sun, for example, transfers its energy to us by way of radiation. Even though radiation amounts to only about 10% of the heat transfer in an engine, it is still important.

Our main interest here is heat transfer in the combustion chamber. Much of the heat is transferred out through the walls of the cylinder. The heat within the cylinder varies from stroke to stroke and overall it is cyclic. Let's consider what happens, beginning with the compression stroke. During the compression stroke the gas temperature increases and there is convective heating of the cylinder walls. The highest temperatures are reached during combustion; they decrease during expansion and exhaust. The cylinder is surrounded by a cooling chamber through which a coolant is circulated. Heat is transferred via conduction through the cylinder walls to the coolant.

Because the temperature within the chamber is cyclic, the heat transfer to the cylinder walls is also cyclic. The cycle times, however, are exceedingly short and therefore the conductive transfer through the walls oscillates for only a very small surface depth. Ninety percent of the oscillations are damped out within a millimeter of the surface; beyond this depth the conduction is steady state. The temperature of the walls inside the combustion chamber might be 190° to 200° C, with the temperature of the coolant being 105° C.

Cooling of the face of the piston is due mainly to convection, with lubricating oil being splashed over the back of the piston face. Conduction occurs through the piston rings to the cylinder walls and also down the connecting rod.

All of this activity, of course, requires a cooling system. The cooling system of a car consists of several parts: a fan and radiator, a water pump, a thermostat, and, of course, coolant to circulate through the system. Water was used as the coolant at one time but pure water has several drawbacks. First of all, it freezes at 0° C, which is unacceptable in northern winter climates. Furthermore, its boiling temperature is lower than desired. I got my radiator repaired several years ago in the spring. Probably because the mechanic thought it was too late in the year for a freeze, he filled the radiator with water instead of returning the antifreeze to it. Needless to say, it was well below freezing the next morning. Not knowing there was water in my radiator I was surprised when the car wouldn't start. I took the radiator cap off and looked inside. Ice?? As you can guess, I didn't take my car back to that mechanic.

Most automobiles now use a 50-50 mixture of ethylene glycol and water (antifreeze). This antifreeze also helps lubricate the water pump and it inhibits rust.

The coolant picks up a lot of heat as it circulates through the motor, and it must therefore be cooled. This is done by the radiator, where the coolant is passed through a large cooling surface. As the coolant runs through the honeycomb channels in the radiator, air rushes through, cooling it. Air flow occurs both as a result of the fan behind the radiator and the forward motion of the car. The water pump is needed to keep the water circulating through the system, and the thermostat is important when the engine is cold. When it is below a certain temperature, no coolant is allowed to flow through the radiator. This helps the engine warm up quickly. As soon as it is warmed up, the thermostat releases the fluid to flow in its usual path.

The Mechanical Sweatshop: Thermodynamics of Engines

The study of the movement of heat and its relationship to work is referred to as thermodynamics, and central to thermodynamics are the PV diagrams mentioned earlier. We saw that the area under the curve in such a diagram gives a measure of the work. In this section we will look at the PV diagram of the internal combustion engine. A detailed analysis of the four cycles gives considerable insight into the engine.

As it turns out, the standard combustion engine

cycle is too complex to analyze precisely, so we make a number of approximations. This isn't all bad, though; even with these approximations the idealized cycle is relatively close to the real cycle. We refer to this cycle as the *air-standard cycle*. It differs from the real cycle in that the gas mixture in the cylinder is treated as if it were air during the entire cycle. Also, the real cycle is an open cycle (valves are opened); we will treat it as a closed cycle. This assumes that the exhaust gases are fed back into the system, and of course they aren't, but this doesn't cause a problem.

Pure air cannot undergo combustion, so we replace the combustion process with the input of heat Q_{in}. Similarly, in the exhaust process we assume it is replaced by Q_{out}. Finally, we assume the engine processes are ideal so that standard formulas can be used (we won't go into them, however).

A final assumption we will make is that the engine has a wide-open throttle. The partially open throttle cycle is slightly different. This idealized air-standard cycle is called the Otto cycle in honor of Nikolaus Otto's invention of the four-cycle engine (fig. 16).

Let's begin with the intake stroke. The piston is at T_C. During this cycle the volume increases from V_T to V_B, shown as $6 \to 1$ in figure 16. The second stroke is the compression stroke, shown as $1 \to 2$. Near the end of this stroke in a real engine, the spark plug fires. This induces a sudden increase in pressure and is represented by $2 \to 3$. The temperature increases dramatically during this stage. In our diagram we assume Q_{in} enters the engine.

The power stroke is shown from $3 \to 4$. During this stroke both the pressure and temperature decrease as the volume increases from V_T to V_B. In the real engine, the exhaust valve is opened near the end of this cycle, and most of the exhaust leaves ($4 \to 5$). In our idealized

Fig. 16. PV diagram for
the Otto cycle.

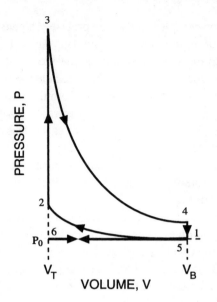

engine we replace this loss with Q_{out}. In the last stroke,
the exhaust stroke, the piston moves from $5 \rightarrow 6$, and
the valve is open so the pressure is constant. The piston is now back at the top to begin the cycle again.

The indicated thermal efficiency is given by

$$E_t = 1 - (1/(V_B/V_T)^4),$$

where V_B/V_T is the compression ratio. Efficiencies given
by this formula will usually turn out to be around 50%.
Brake thermal efficiencies are usually about 30%. And
finally, the work done during the cycle is the area inside
the curve.

The Carnot Cycle

We saw earlier that considerable heat goes out through
the exhaust system. This is, of course, wasted energy,

as heat is a form of energy. Is it possible to minimize these heat losses and make our engine more efficient? It would be great if we could use all the heat produced by the engine, but that is impossible. No one has ever produced a heat engine that does not throw away a considerable fraction of the heat supplied to it. Indeed, according to one of the fundamental laws of thermodynamics, known as the second law of thermodynamics, no one ever will. In case you're worried that we skipped the first law, it only states the law of conservation of energy (i.e., no energy can be gained or lost in a closed system). The second law states that all the heat in an engine cannot be converted to energy. Yet strangely, there is no problem with the reverse process: all the energy of a system can be converted to heat.

A young French engineer, Sadi Carnot, was interested in how much work a heat engine could do. Carnot's main interest was steam engines, but the principles he developed apply to all heat engines, including the internal combustion engine. Steam engines at that time, in the early 1800s, were notoriously inefficient (about 5% efficient). This meant that approximately 95% of the heat energy of the burning fuel was being wasted. Carnot was determined to find out how this efficiency could be improved. He began by disregarding the details of the heat engine and focusing only on the fact that the engine was supplied with energy in the form of heat at a high temperature (call it T_2). The engine performs work and rejects heat at a lower temperature T_1.

Carnot was interested in maximizing the efficiency of the engine given that the heat was supplied at temperature T_2 and heat was given off to the exhaust at T_1. He visualized what is now called a Carnot engine. We can draw a PV diagram of it as we did previously (fig. 17). He showed that the Carnot cycle was the most efficient heat engine. Of particular importance is that it is bound

Fig. 17. PV diagram for
the Carnot cycle.

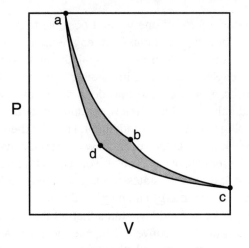

by two isothermal curves (with the same temperature throughout). Thus the heat is supplied at a single high temperature and exits at a single low temperature.

It is easy to show that no engine operating between the temperatures T_1 and T_2 can be more efficient than a Carnot engine. The Carnot engine is, of course, an idealized engine in which there are no losses due to friction or radiation heat. What is its efficiency? If we use absolute (kelvin) temperatures, we can write it as

$$E_{carnot} = 1 - T_1/T_2.$$

This is the maximum efficiency a heat engine can attain. It's easy to see from this equation that if we want to maximize the efficiency we have to make T_1, the exhaust temperature, as low as possible, and T_2 the engine temperature, as high as possible. The only way we could get 100% efficiency is to make $T_1 = 0$ K, and this is impossible by the third law of thermodynamics.

In practice, of course, no internal combustion engines approach the efficiency of a Carnot engine. There are all kinds of energy losses.

The Diesel Engine

So far we have been talking only about the internal combustion engine, but there is another engine that is commonly used in both trucks and cars: the Diesel engine. The main difference between the diesel and the spark-ignited engine is that the diesel does not require a spark plug.

This engine was the brainchild of Rudolf Christian Diesel. As a young man, Diesel saw a glass-cylinder version of the ancient Chinese "firestick." One would place a small piece of tinder into the cylinder, and then a rapid stroke of the plunger would ignite it under the heat of compression. To Diesel this process was miraculous, and it left an indelible impression on him.

Carnot's book on thermodynamics and engines was Diesel's bible and he studied it diligently. His objective was to increase the efficiency of the steam engine (the spark-ignited combustion engine was only in its developmental stage at the time). After considerable experimentation he showed that a fuel-air mixture, when compressed to a high pressure, would ignite and combust. It was used as the basis of his four-cycle "diesel engine."

The four-stroke diesel engine is quite similar to the four-stroke spark-ignition engine. In the intake stroke, the piston moves toward B_c and the intake valve opens with air being drawn into the cylinder. The intake and exhaust valves are then closed and the piston moves upward, compressing the air. Just before it reaches ignition, fuel injectors spray fuel into the cylinder. The fuel combusts and the piston moves downward, giving the power stroke. In the final stroke, the exhaust stroke, the exhaust valve is opened and the exhaust gases are pushed out.

Diesel engines run on diesel fuel rather than gasoline

because it is more efficient: it contains about 10% more energy per gallon than gasoline. Turbocharging of diesels is particularly desirable as it gives much better fuel efficiency and more power.

The Rotary Engine

The piston engine has a number of disadvantages and many engineers have tried to replace it with a more efficient engine. One of the most successful of these alternatives is the rotary engine, the best known of which is the Wankel engine, invented in the mid-1950s by Felix Wankel of Germany. It has been used by the Japanese in the Mazda models.

The Wankel engine uses a three-cornered rotor that turns on an eccentric gear (see fig. 18). The rotor divides the main chamber into three sections, and at any given time the usual features of the piston engine—namely, intake, compression, power, and exhaust—are taking place in a given section of the chamber. As with the piston engine, the fuel-air mixture is brought in and compressed as the rotor rotates. It is then ignited by a spark

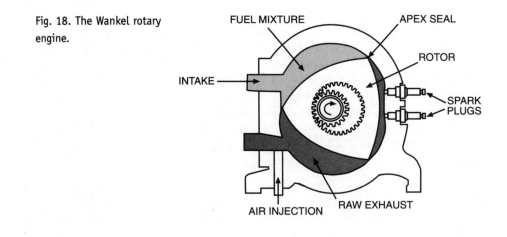

Fig. 18. The Wankel rotary engine.

plug, which produces the turning power, and finally the exhaust gases are pushed out.

The major advantage of the rotary engine is that it doesn't have cylinders, pistons, valves, and a crankshaft. All these are done away with, and therefore the engine has fewer moving parts and can be smaller and lighter. Another advantage of the Wankel engine is that it generates a power impulse with each revolution rather than through the two cycles of the piston engine. Although a rotary engine can provide the power of a piston engine that weighs twice as much, it does have disadvantages: it uses about 20% more fuel than conventional engines, and it is a high polluter.

The W Engine

Volkswagen has recently introduced a W engine. Why would anyone want to make an engine in the shape of a W? It does, indeed, have some advantages. It is more compact than a similar V8, and it can produce a lot of power. The easiest way to think of the engine is as two V4 engines side-by-side. The W engine has four banks of two cylinders and is about the same width as a V8 but considerably shorter. The engines produced so far have a respectable 275 hp at 6000 rpm and a torque of 273 at 2750 rpm.

More powerful models are planned for the next few years. Audi is planning a W12, which will have three cylinders per bank. It will have a horsepower of 414 at 6000 rpm and a torque of 406 at 3500 rpm. Earlier I mentioned the Bugatti EB 16. It is scheduled to have a W engine and, as we saw earlier, it will have a horsepower of 987.

As this chapter has shown, physics is important in relation to the internal combustion engine and its characteristics. Indeed, it is indispensable.

CHAPTER 4

When Sparks Fly

The Electrical System

Late one night on our way home from skiing, our car suddenly stopped. Because it was a relatively new car, I was surprised that there was a problem. I soon realized it was in the electrical system so I checked the wiring, the condenser, and points, then took the rotor out and examined it closely with a flashlight. Nothing appeared to be wrong. But the car still wouldn't start, so we had it towed to a nearby garage and I watched as the mechanic went over everything. He seemed a little confused at first, then finally smiled. "That's it," he said, holding up the rotor. "It's got a fine-line crack in it . . . you can barely see it." I looked at it and, sure enough, he was right.

Cars no longer have rotors so I don't have to worry about this particular problem ever happening again. Nevertheless, there are still things that can go wrong with the electrical system. Furthermore, this system has become much more complicated. Unless you're a master mechanic, the maze of wires and electronics under the hood of a modern car look almost as complicated to you as the inside of a computer. Indeed, many of these wires are actually hooked to a computer, since computers are now an integral part of cars. We won't go into detail on computers, but I will try to impart a sense of how the electrical system of a car works.

Let's begin with a little background. As you probably know, the main component of electricity is the electron, or rather, electrons in motion. An electrical current is, in fact, nothing more than a stream of electrons moving along a copper wire. Why copper? To answer this we have to look at the atom. An atom is composed of a positive nucleus, surrounded by negatively charged electrons in various orbits, or shells. The electrons are held in place because positive charges attract negative charges, and there are equal numbers of each. What we're interested in is the outermost shell of electrons, known as the valence shell. In some materials, called conductors, one or two of the electrons in this shell are barely held to the atom; in fact, they are able to wander back and forth between neighboring atoms. In a sense, this "wandering" constitutes a current, but it is random and because of this it isn't significant. If we apply a positive charge to one end of the copper material, however, the electrons will be attracted to it. They will jump from atom to atom in the direction of the positive charge. In reality what we are doing is applying a *voltage* between the ends of the conductor.

Voltage is a "pressure" that is put on the electrons to move them in a certain direction. In many ways it is like the pressure that is applied to water to get it flowing. Indeed, as we will see, there are many similarities between flowing water and the flow of electricity, and I will discuss several of them as we go through the chapter.

Voltage can be either positive or negative. This is referred to as its polarity, and it determines the direction the electrons will flow in the wire. When a voltage, or more exactly, a voltage difference, is applied to a wire, electrons will flow and we will have a *current*. Electrical current is like the current you get in the flow of water, and just as we measure water flow as so many gallons per second, we also need a measure of the number of electrons that are flowing in the wire. It might seem that a convenient measure would be the number of electrons flowing per second, but when you look into this you see that this number is usually several billion billion. We therefore use the unit called the *ampere,* which is about 6 billion billion electrons per second.

What do electrons do as they flow around a circuit? Their major function, of course, is work. They turn over a motor, light a lamp, or supply heat as well as many other things. A question that immediately comes to mind is: How do we measure this work? It obviously involves both voltage and current, so it must be related to them. Actually, what we would prefer is the rate of doing work, or power, and we do have a unit of power. As we saw earlier, it is called the *watt* (or kilowatt, which is a thousand watts). Power is obtained by multiplying current by voltage.

Because power is dependent on both current and voltage, both are important in performing work. High voltage with almost no current doesn't give much power, and the same goes for high current and little voltage. This reminds me of a question some people once asked me: What really kills you—the high voltage or the high current? The couple was having an argument about this and wanted me to resolve it. I told them it takes both. Extremely high current with little voltage will not harm you, and very high voltage with almost no current will give you a pretty good tingle, but it won't kill you.

I had an interesting experience in relation to this a few years ago. I was giving a lecture on electricity, demonstrating various things about it using a Van de Graaff generator—a device for generating a large charge. I was showing the students how a fluorescent bulb would light up when you held it near the charged ball at the top of the Van de Graaff. At the beginning of the lecture I had told a couple of jokes about electricity and barely got a chuckle, much to my disappointment. As I talked about the Van de Graaff I pointed toward it. All of a sudden a bolt of lightning arced out of it and struck the end of my finger. I'm sure I jumped several inches off the floor. There was a brief silence, then the class started to laugh, and they continued laughing so long it was hard to stop them. I had got my laugh after all, but not in the way I expected.

Getting back to voltage and current, there is something else that is important in electricity, and it's called *resistance*. Let's go back to our analogy with flowing water. If the water in a pipe encounters a restriction, or sudden narrowing of the pipe, the flow rate will be reduced. The same goes for electrical circuits; if there is a restriction, the rate of flow, or current, will change. This opposition to the flow of electrical current is called resistance. All circuits have resistance of one kind or another in them. It is measured in units of ohms (designated by Ω).

You have now been introduced to the three major components of electricity: voltage, current, and resistance. It's natural to ask if there is a relation between them. And indeed there is. It is called Ohm's law. We write it as volts = current × resistance, or

$$V = IR.$$

If you want the current in a circuit, you merely write this as $I = V/R$. This relationship is particularly useful for determining currents and voltages in a circuit, and I will give a number of examples using it later.

We saw earlier that we need the wire in our circuit

to be a good conductor. Most wires are made of copper, which of course is one of the best conductors. Besides conductors, however, there are insulators, which have few free electrons and are very resistant to the flow of current. Also, in between conductors and insulators are semiconductors, which are important in relation to transistors and diodes. It might seem that insulators are of little use in electrical circuits, but they are important, being needed to isolate wires from one another. Without a coating of insulating material around them there would be a large number of "shorts" between the wires. Indeed, a wire such as a spark plug wire needs considerable insulation because of its high voltage.

Circuits

In any car there are a large number of circuits. The starter, ignition system, charging system, lighting system, and accessories such as the radio all need electricity. In any given circuit there are three major components: a battery, or source of voltage, conducting wires, and a load (see fig. 19).

One of the first things we have to consider in such a circuit is: What is the direction of the current? We saw earlier that the electrons are attracted toward a positive charge; therefore they move from the negative pole of

Fig. 19. Simple series circuit showing source of voltage and load (resistor).

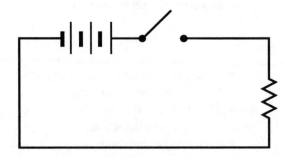

the battery, through the load, to the positive pole. Due to an accident of history, however, we do not consider this to be the direction of the current. The conventional current direction is taken as positive to negative, opposite the flow of electrons. This sounds a little crazy, but it's not as bad as you might think, and, as it has turned out, it is convenient. First of all, we usually ground the negative terminal of the battery, so it makes sense to think of the current as coming from the positive pole. Second, we will later talk about transistors and semiconductors, where we have positive charges called "holes" flowing in the circuit. In this case, conventional current direction is just the direction of flow of the positive charges. We will use conventional current direction throughout the book.

Let's begin with the simplest type of circuit, the *series circuit* (fig. 20). Many of the circuits in a car are of this type. The resistor shown in figure 20 could represent many things, such as a lightbulb or an actual resistor (each is important in circuits). We'll assume in this case that the resistance is 6 ohms. The battery is assumed to be 12 volts, since that is the usual voltage of a car

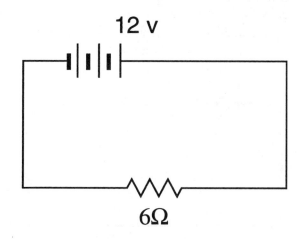

Fig. 20. Series circuit with 12V battery and 6Ω resistor.

battery. Having this information we can use Ohm's law to calculate the current that will flow in the circuit. It is 12 volts/6 ohms = 2 amps. Note that if you had several resistors in the above circuit you could easily find the total resistance. It is merely the sum of the resistances.

We turn our attention now to the voltage drop in the circuit. If we put a voltmeter across the 6-ohm resistor we will see a voltage drop of 12 volts. What this tells us is that the 12 volts of the battery are all used up by the resistor. On the other hand, if we had several resistors in series—say, the 6-ohm resistor was made up of three 2-ohm resistors, then the voltage drop across one of them would be 2 amps × 2 ohms = 4 volts. But the total across the three would again have to be 12 volts.

It doesn't take much know-how to realize that series circuits have several disadvantages. As more resistance is added, the current goes down and the voltage drop across each load becomes less. This could cause problems in a circuit where a particular voltage or current is needed. Also, if the two resistors in a circuit were two bulbs, say the headlights, and one burned out, the other would also go out because of the short circuit in the line. This isn't something we want.

We get around problems such as this by using *parallel circuits*. In this case the resistances are in parallel, as shown in figure 21. If one of the bulbs burns out, the

Fig. 21. Simple parallel circuit.

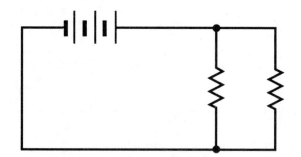

other will remain lit, since the current through it is still closed. Also, of particular importance, the voltage across the two resistors is the same, which is necessary when two bulbs require the same voltage to operate. In our example, however, note that less current will be going through the bulbs than in the series circuit. Remember the analogy to water: if a water pipe splits and becomes two, there's going to be less water in one of the pipes than there was in the original one.

We can still use Ohm's law to make calculations here. Suppose the voltage across each of the resistors (bulbs) is 12 volts. With the resistance on the left being 3 ohms, the current through it is 12/3 = 4 amps (fig. 22). The current through the other bulb is 12/4 = 3 amps. What about the current in the main circuit? We obviously have to add the two resistors to get it, but we can't do it in the way we did with series circuits. Resistors in parallel add differently; they add according to the formula

$$R = 1/(1/R_1 + 1/R_2),$$

where R_1 and R_2 are the two resistors in parallel. In the above case, $R = 1/(1/3 + 1/4) = 1.72$ ohms. Thus, the resistance of the overall circuit is 1.72 ohms, so the current that flows here is 12/1.72 = 6.97 amps. Finally, it is

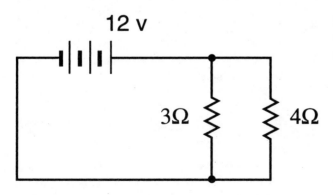

12 v

3Ω   4Ω

Fig. 22. Parallel circuit with 3Ω and 4Ω resistances in parallel.

Fig. 23. Circuit with
both series and parallel
resistors.

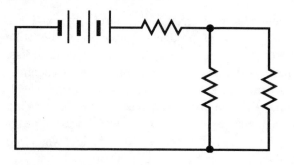

possible to have combinations of series and parallel circuits such as the one in figure 23, but they are not used very often in cars.

It is clear from the circuit diagrams we have seen so far that a closed or complete circuit is needed for current flow. If there is a break in the circuit, you have an open circuit, and no current flows. But in cars, complete circuits take a lot of wire, and we can cut down on this by using a *ground*. The simplest way to do this is to use the frame as part of the circuit (fig. 24). We represent this in a circuit diagram as in figure 25. The symbol ⏚ represents a ground. An electric fence works in the same way, but there you are actually using the ground to complete the circuit. Electric fences are, indeed, amazing things: they work even when they aren't plugged in. Once the animals inside the fence have touched the fence, they won't go near it again for months.

Incidentally, we can summarize everything about series and parallel circuits in two sets of rules that are helpful to remember.

Series
1. The current throughout a series circuit is the same.
2. The total voltage drop across the loads is equal to the battery voltage.

3. The voltage drop across each load depends on the resistance and can be obtained from Ohm's law.
4. The total resistance of the circuit is the sum of the resistances.

Parallel
1. The current through each branch is different and can be calculated from Ohm's law.
2. The voltage across each load is the same, being the battery voltage.
3. The current in the main circuit is the sum of the currents in the branches.

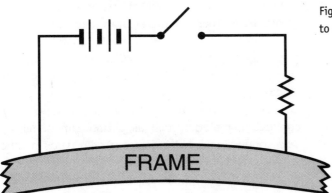

Fig. 24. Circuit grounded to the frame.

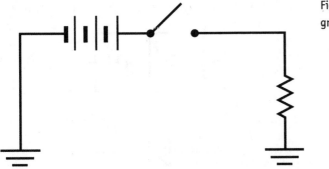

Fig. 25. Circuit showing grounds.

4. The total resistance of the branch resistors is less than the lowest resistance in the branches.

Transistors and Diodes

So far we have encountered resistors and batteries in circuits. We will see that motors, alternators, relays, and so forth are also found in car circuits. For many years, electronic components—transistors and diodes—were used in cars. Today they have for the most part been replaced by thin pieces of silicon containing numerous electronic parts, or integrated circuits. Most electronic parts in the circuit, however, are tiny transistors. Indeed, there are sometimes as many as 10,000 transistors on a tiny integrated circuit. These circuits are now referred to as chips. With cars becoming largely computer controlled, chips play a significant role in them. We won't say much about chips, but it is important to understand transistors and diodes.

Let's begin with diodes. Diodes are made of semiconducting material such as germanium or silicon. Semiconductors can be doped p or n type; in other words, positive or negative according to the type of charges. A diode is made up of a p and an n type semiconductor (fig. 26). This combination allows current to flow in only one direction. In circuits this is represented by —▷|—, where the arrow shows the direction of current

Fig. 26. A p-n junction and its circuit representation.

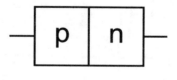

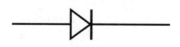

flow. One of the most important uses of diodes is to rectify the current from an alternator. As we will see later, the current that comes out of an alternator is AC (alternating), but we need DC (direct, or uniform) current to charge the battery and to run other things in the car. The AC current can be changed to DC using diodes. Diodes are also used to protect and insulate vulnerable circuits from voltage or current surges, particularly those that occur near coils.

Even more important in most electronic circuits are transistors. In practice, the transistor isn't much more complex than the diode; we merely add another n or p section to a diode to get it. There are two types of transistors: npn and pnp. Note that there are also three connections. In a circuit we represent them as shown in fig. 27.

It is convenient to think of transistors as taps, where the base is the tap handle that controls the water flow. In an electrical circuit the base controls the current through the two sections we call the emitter and the collector. When you adjust or change the current in the base, it affects the current going from the emitter to

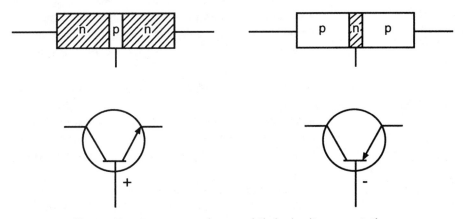

Fig. 27. Transistors npn and pnp and their circuit representations.

the collector. This means the transistor is like an amplifier, where a small signal controls a large signal.

With cars becoming more and more electronic and computerized, transistors are widely used. In particular, they are used in the ignition circuit, the charging circuit, and the starter circuit. Actually, as I mentioned earlier, we now use integrated circuits, but transistors are an important component of these circuits.

The Battery

The battery is an essential component of a car: a car won't run without a battery. It also won't run if the battery is dead, or at least you won't get the engine started. Most people have had an experience with a dead battery. When I was going to college, several students rode with me each morning to campus. My battery was on its last leg, but as I was always short on money I kept putting off buying a new one. It had roughly enough juice in it each morning to turn the engine over two or three times. If it didn't start, we could push the car to get it going. After several mornings of this I decided it was time to buy a new battery. As I walked up to my car and my waiting passengers, I was sure they'd be wondering whether it would start. As I got close to them, I saw that they had a surprise for me on the grass: a brand-new battery.

There are two types of batteries: chargeable and unchargeable. The small batteries that you put in your flashlight are usually unchargeable. The battery in your car, on the other hand, is chargeable, and it's the only type I'll talk about. As we saw earlier, most car batteries are 12 volt.

When you charge a battery you aren't actually storing electricity in it. What you are doing is changing the chemicals within the battery to make it operable again,

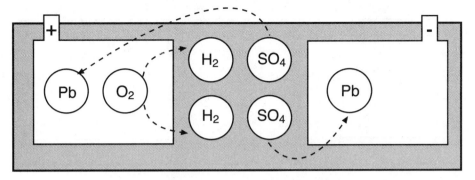

Fig. 28. Chemical processes in a discharging battery.

or bringing it up to maximum operating conditions. Let's begin with a typical battery cell. It's made up of a positive and a negative plate, separated by a separator, and an electrolyte. A plate is made up of a screenlike grid that is composed of antimony or calcium. Lead or lead oxide is pressed into it, depending on whether you want a positive or negative plate. The plates are separated by insulators of plastic or glass. The electrolyte, or the solution in which the plates and insulators are placed, is a mixture of sulfuric acid (H_2SO_4) and water. In a given cell, many plates are used and all the positive and all the negative plates are wired together in series.

In a fully charged battery we have positive plates composed mainly of lead oxide (PbO_2) and negative plates composed mainly of spongy lead (Pb). When you start the car or turn on the lights, the battery discharges. Let's consider what happens. In this case, oxygen (O_2) from the lead oxide breaks free from the positive plate and joins with the H_2 in the solution. The SO_4 from the breakdown of sulfuric acid combines with the lead in the negative plate, and finally the SO_4 from the solution combines with the lead in the positive plate. If the

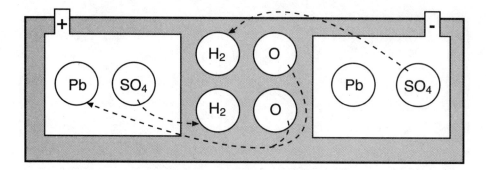

Fig. 29. Chemical processes in a charging battery.

process of discharging continues long enough, the two plates will end up as $PbSO_4$ (fig. 28). With the two plates being the same, there is no potential difference between them, and the battery is dead.

To recharge the battery we have to reverse the process; in other words, we have to make the positive plate PbO_2 and the negative plate Pb. For this, SO_4 from the positive plate must form with H_2 in solution. Oxygen from the solution must join with Pb in the positive plate, and SO_4 from the negative plate must leave and join H_2 in solution to form sulfuric acid (fig. 29).

Starting the Car

The first thing you do when you jump in a car is start the engine. To you, it's as simple as turning the key, but what goes on within the car and electrical system is not that simple. Let's take a look. We'll begin with the starter motor. The physics behind the motor involves the interaction of currents and magnetic fields. As we saw earlier, we can produce a magnetic field using a current. Indeed, all wires carrying a current have a magnetic field associated with them. You can determine the direction of this field (magnetic field lines run north to south) if you grasp the wire in your right hand and point your thumb in the direction of the current. The field will be in the direction of your fingers.

If several loops of wire are wound around an iron core and current is passed through them, you have an electromagnet. The iron enhances the magnetic field. The first electromagnet was built by the English experimenter William Sturgeon. In 1823 he wrapped eighteen turns of bare copper wire around a U-shaped bar and was delighted when it lifted twenty times its weight in iron. A few years later, in 1829, the American physicist Joseph Henry picked up on the discovery. Using insulated wire he wound hundreds of turns around an iron core and produced a particularly strong electromagnet. In 1833 he made one that could lift a ton of iron.

Electromagnets are found in many parts of a car. One of the reasons they're used so much is that magnetic fields interact with one another. Two wires with current in them running in the same direction will be attracted to each other because of these interacting fields; if the currents are in the opposite direction, they will repel each other. On the basis of this knowledge, let's look at what happens when a wire with current passing through it is placed in a magnetic field. The magnetic field of the wire will interact with the magnetic field of the magnet. On one side of the wire the field will be enhanced, on the other side the fields will cancel. The result will be a strong field on one side and a weak field on the other, and since field lines act like stretched rubber bands, the strong ones will tend to push the wire shown in figure 30 to the right.

Now let's bend the wire as shown in figure 31. The loop of wire will be twisted by the magnetic field. But, after half a turn there will be no force on it. If we can somewhere reverse the current through the loop, however, the force will continue twisting the loop. We can

Fig. 30. Wire with a current passing through it in a magnetic field. The interaction of the two magnetic fields (wire and permanent magnet) tends to push the wire out of the permanent magnet's field.

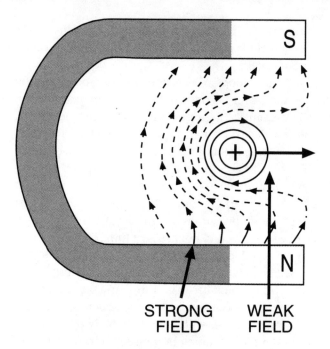

STRONG FIELD **WEAK FIELD**

do this by introducing what are called commutator and brushes (fig. 32). The commutator is shaped like half circles, and we can enhance its effect by adding many loops of wire. With enough loops of wire we will get an armature, which is the main component of a motor.

Fig. 31. Loop of wire in a magnetic field (end view). The field forces the wire to rotate.

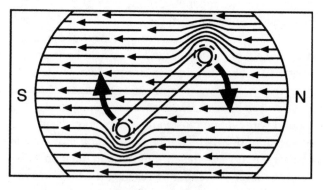

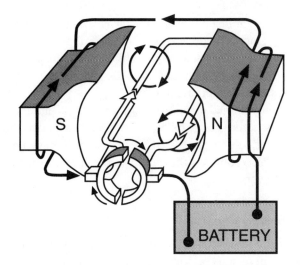

Fig. 32. Loop of wire in a magnetic field connected to a commutator that reverses the current.

S

N

BATTERY

So, we now have a motor, which is turned over when we turn the key and press the starter (see fig. 33). At one end of this motor we have the starter drive, a small ring gear that engages a larger gear on the flywheel. When the flywheel is spun, the pistons move up and down and the spark plugs fire, starting the engine. It's important

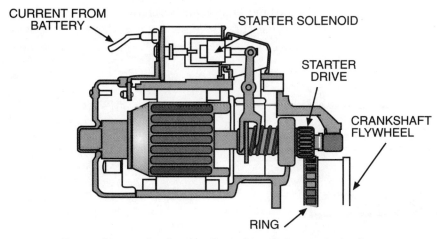

CURRENT FROM BATTERY

STARTER SOLENOID

STARTER DRIVE

CRANKSHAFT FLYWHEEL

RING

Fig. 33. Starter, showing the starter drive that engages the ring gear.

that this gear disengage as quickly as possible after the engine starts; if it doesn't, considerable damage to the motor and gears could result. This is accomplished by an overrunning clutch. Basically, the clutch is one-way, allowing motion in one direction only.

Finally, I should mention the starter solenoid, mounted on the starter motor. Its main function is to move the starter drive gear so that it meshes with the flywheel gear. It also connects the battery to the starter motor to turn it over.

The Charging System

It takes a lot of current to start your car and therefore your battery must be constantly recharged. In fact, if you have trouble starting your car, you can easily end up with a dead battery trying to get it going. In short, anything you take out of the battery must be put back in: it has to be recharged. Furthermore, you need current to run your car, and you get it from the alternator. If you're from the older generation you might say, "Don't you mean the generator?" Indeed, DC generators were common in cars at one time, but all cars now have AC alternators.

So, let's turn to the alternator. Earlier we saw that we could cause a wire to move if we placed it in a magnetic

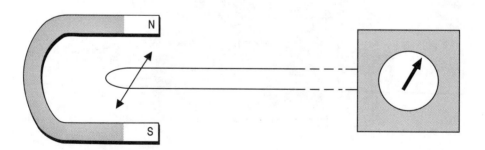

Fig. 34. A moving wire in a magnetic field creates a current within it.

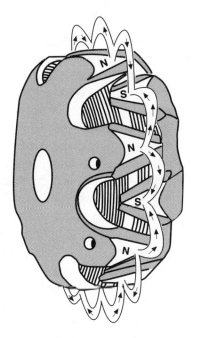

Fig. 35. Alternator, showing rotor and stator. The rotor and stator are both coils of wire. The rotor moves; the stator, which surrounds it, does not.

field and passed a current through it. But what happens if we reverse things? Assume the wire is in a magnetic field with no current passing through it and we move it. You will notice that a current flows—in other words, a voltage is induced in the wire (fig. 34). This means we can generate electricity if we mechanically move a wire in a magnetic field, which is what an alternator does.

An alternator consists of two main parts: a rotor and a stator (fig. 35). As we saw, we need a coil of wire in a moving magnetic field, or a moving coil of wire in a stationary magnetic field. It turns out that it is more convenient to move the magnetic field. The moving field is the rotor; it consists of a coil of wire that produces a magnetic field, and it is placed between two sets of interlinking fingers as shown in figure 35. The fingers will be north or south, depending on whether or where they are relative to the coil. They are placed so that there are alternating north and south poles. The current is brought into the rotor through slip rings, similar to the commutator in the motor. The rotor is made to rotate by hooking it to the engine using a pulley.

The stator is a case that contains many coils of wire that surround the rotor. As the rotor turns, its magnetic field lines cross the windings of the stator and induce a current. The windings of the stator will be exposed alternately to north and south magnetic poles. The current that is generated will therefore run back and forth in the windings; in other words, it will be AC. But we don't want AC. To recharge the battery we need DC. Furthermore, to operate most systems in the car we need DC. How do we get it? This is where the diodes we talked about earlier come in. We can use them to rectify the AC. (See fig. 36.)

We know that a diode allows current to flow in only one direction. Let's see how this affects AC. As figure 36 shows, the diode blocks off any current in the negative direction so we have half-loops with nothing between them. But if a diode is hooked up in the opposite direction it will block off alternate loops. With the appropriate combination of diodes we get the current shown at the bottom in figure 36. And from here it is straight forward to DC.

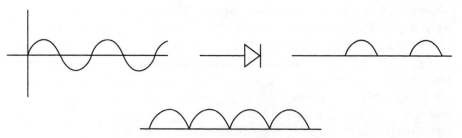

Fig. 36. Conversion of AC to DC via a diode.

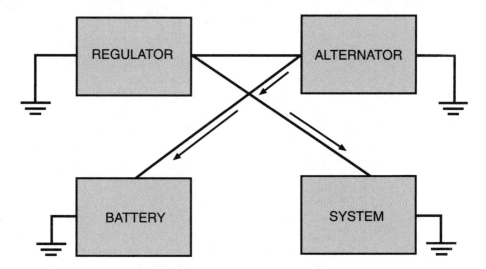

Fig. 37. Regulatory system of a car.

Now that we have DC we still have a problem. The current that we put back in the battery and use in the car has to be carefully regulated. If the charging system of the car is operating at 11 volts and the battery is 12 volts, it will soon run down. We must direct current to the battery when it is low, then when it is fully charged we must stop it. We don't want to overcharge it.

How do we control, or regulate, the output of the alternator? Looking back at the rotor and stator we see that it is the current of the coil inside the rotor that is critical. The current going through this coil is called the *field current.* If you increase this field current, the output from the stator will increase, and it will decrease if the field current is decreased. What we need, then, is a variable resistor in the field coil current loop. When we change this resistor, the output will change.

To be effective the regulator must be able to sense the voltage and current in various parts of the system.

In particular, it must be able to sense a voltage drop such as the one that occurs when the headlights are turned on. Regulators have become computerized in recent years and are now part of the car's overall computer. (See fig. 37.)

The Ignition System

One of the main jobs of the electrical system is to supply electricity to the spark plugs so that they can combust the fuel-air mixture in the cylinder and drive the pistons. The voltage needed for this can be 20,000 volts or more, and the battery only supplies 12 volts. The needed voltage comes from the ignition coil. And as we will see, the ignition coil is based on another principle of physics that involves the interplay between currents and magnetic fields.

We saw that if we wind wire around a core of soft iron we get a strong magnetic field. Let's start with that. Assume we make a few hundred turns around the core with relatively heavy wire; this becomes the primary circuit. Now take much finer wire and wind thousands of turns over the primary circuit. This is called the secondary circuit. (See fig. 38.) There will be much more resistance in the secondary compared to the primary circuit because there is so much more wire. Now, arrange to have the primary circuit turned off and on quickly. At one time this was done with points, but today an electronics device is used. Consider what happens. When the current is turned on, the magnetic field builds up; then as the primary circuit is opened, the magnetic field collapses. In particular, it moves across or cuts the secondary windings, inducing a high voltage in the secondary circuit. The ratio of the voltages will go as the ratio of the number of windings in the primary and secondary circuits, so we can get 20,000 volts out

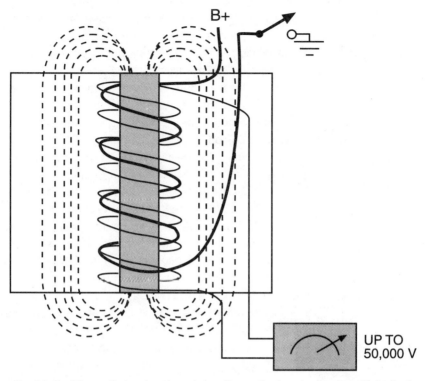

B+

UP TO
50,000 V

Fig. 38. Ignition coil showing the primary (heavy line) and secondary (light line).

of the secondary without too much trouble. This is the voltage that is applied to the spark plug. The current will, of course, be very low.

The ignition coil is nothing more than a transformer—which was one of my favorite devices when I was in high school. I built several, hoping to use them to build a high-voltage "lightning maker" (like the Van de Graaff I talked about earlier). After a couple of frightening experiences, however, I decided to abandon the project.

The high voltage has to be applied at just the right time—in other words, when it is needed. In older systems the output of the coil was directed to the spark plugs by way of the distributor. Because modern cars no longer have distributors and they are becoming a thing of the past, I won't discuss them here. They have been replaced by electronics and sensors.

Since we have both extremely high and low voltages in the overall ignition system, we have a primary circuit and a secondary circuit. The primary circuit is the low-voltage control circuit; the secondary circuit is the high-voltage circuit to the spark plugs. The overall circuit will look similar to the circuit shown in figure 39. What is going on here is the following. A sensor senses the position of the piston so that it can determine when a spark is need by the spark plug. Information is forwarded to the primary circuit, which in turn produces a high voltage impulse in the secondary. This high-voltage impulse travels along the secondary circuit to the spark plugs, firing them and combusting the fuel.

There are several different types of piston sensors or crankshaft position sensors. For the most part they use small magnetic fields. Small magnets, for example, can be placed on the crankshaft, and the sensor is a coil of

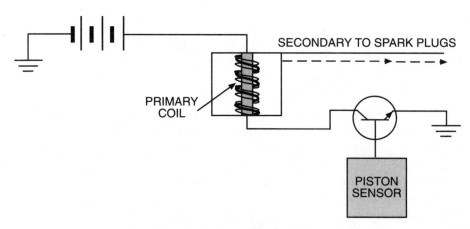

Fig. 39. Ignition system circuit.

wire. When a magnet passes the coil, it generates a small current in it. Slightly different, but based on the same principle, are the commonly used Hall effect sensors.

As we have seen, electricity and magnetism—two very important branches of physics—are also important in relation to cars. Without them cars could not exist.

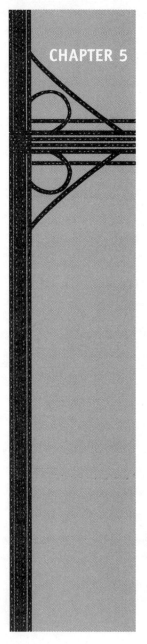

CHAPTER 5

"Give 'em a Brake"

Slowing Down

Almost everyone will agree that good brakes are vital. If an emergency arose and you had to stop quickly only to find that the brake pedal went to the floor, it would be a frightening experience. I've never had this happen but something similar did occur many years ago, shortly after I got my first car, an older model. I brought it home, beaming with pride. I knew it needed a lot of work, but it was mine. After getting it home I was anxious to try it out, so along with a friend I took it out on the highway. As teenagers, our major interest was in how fast it would go. I was pleased when it got up to 60, and was soon pushing it to see how much faster it would go. I was so intent on increasing its speed that I wasn't watching the road signs as closely as I should have been. Suddenly I realized we were coming up on a sharp turn, and I jammed on the brakes. To my astonishment the car jerked violently to the right. It scared me so much I almost went off the road. I touched the brakes again, this time more lightly, and again they pulled to the right.

We were still doing close to 50, and the sharp turn

was almost upon us. I pumped away at the brake and somehow we made it around the corner safely. I sighed in relief as we got the car to slow down. After this experience I swore I would always check the brakes in a strange car before I drove it.

Modern brakes are a far cry from the brakes of older cars. Hydraulic brakes were introduced in the mid-1920s and were a great innovation. Dual brakes were introduced in 1968, making cars even safer—if one set of brakes gave out, there was always a backup. Most cars now have power brakes, and in recent years we have had the introduction of ABS, or antilocking brakes, which make cars even safer.

Friction

Most people have a pretty good idea what friction is and what it is needed for, particularly during the winter when there's lots of snow and ice on the road and sidewalks. It's friction that stops you from falling on your behind, and it's friction that stops your car when you press the brakes.

When one surface slides over another there can be considerable friction or only a small amount, depending on the roughness or smoothness of the surfaces. You may think that there is almost no friction, say, between a skate blade and ice, but even then there is a small amount.

What causes friction? If you look very closely at any surface you will see irregularities. You may sometimes have to use a microscope to see them, but they are still there. As two surfaces flow over each other, the irregularities in one of them catch those in the other, and the result is a frictional force that opposes the motion. In short, each body exerts a force on the other, and the force is parallel to the surfaces. Frictional forces can also exist when there is no motion between the surfaces. The force on a body, for example, might be insufficient to move it.

Let's try an experiment. Assume you have an object sitting on a surface. Given that this is a book on cars, you can assume it is a wheel and tire. Also assume that you have a spring scale (even though it's not very accurate, for our purposes it will do). If you push on the tire slightly it probably won't move because of the friction between it and the surface it is sitting on. Now, attach the spring to it and pull. If you pull hard enough it will eventually move: you have, in effect, overcome the frictional force resisting the motion. Actually there still is a force, so you haven't completely overcome it, otherwise it would continue moving indefinitely without any exertion on your part.

The friction force between the surfaces when the tire is at rest is called the *static* frictional force. Once the tire begins to move, however, the force decreases and it is referred to as a *kinetic* frictional force. One of the things you see immediately is that the frictional force is independent of the area of contact and proportional only to the force holding the surfaces together. We refer to this force as the *normal* force and usually designate it as N.

Since the frictional force depends on N, we have a mathematical relation between them: $F \leq \mu N$, where μ is the coefficient of friction, which gives us a measure of how rough or smooth the surfaces of contact are or how easily they slide. We use the symbol $\leq$ because it is a variable force. In other words, you can push on the tire with various forces before it moves.

But we must be careful. Assume that we are pulling on the tire but it is not yet moving. In this case we are dealing with static friction and the relationship is $F \leq \mu_s N$ where μ_s is the coefficient of static friction. Once the tire begins to slide, however, we don't have a variability.

One and only one force is required to move it uniformly. Furthermore, the coefficient of friction becomes less once it begins to move, so we can no longer use μ_s. Our new coefficient of friction is μ_k, and we call it the kinetic coefficient friction. Our expression is then

$$F = \mu_k N.$$

Note that both F and N are forces with the same units, therefore, μ_s and μ_k are dimensionless; in other words, they have no units and are a number between 0 and 1. A coefficient of friction near 0 specifies a very slippery surface; near 1, it specifies a rough surface.

A few examples of frictional coefficients are as follows:

	μ_s	μ_k
Rubber on concrete	.90	.70
Copper on glass	.68	.53
Oak wood on oak wood	.54	.32
Brass on steel	.54	.32
Steel on ice	.02	.01

Since we are mainly interested in cars, let's consider several cases for tires on a concrete road:

	μ_s	μ_k
Dry concrete—low speed	.9	.7
Dry concrete—high speed	.6	.4
Wet concrete—low speed	.7	.5

Stopping the Car—Deceleration

When most people think of a car, particularly a high-powered modern car or a race car, they think of acceleration.

How fast does it go from 0 to 60? Indeed, if you look at the specifications of new cars in any car magazine, this is one of the first bits of information provided.

But equally important is: How fast can the car be stopped? As we will see there are limitations—even if the car has excellent brakes—one of the most important of which relates to the force on your body when you stop a car suddenly. (You also experience this force when you accelerate.) As you know, an astronaut experiences several g's when he or she takes off in a rocket. It only stands to reason, therefore, that people driving a car would also experience g's when they accelerate or decelerate.

Let's see what effect g's have on us. If you're decelerating at 1 g you are decelerating at 32 feet per second each second. In other words, if you started at 64 ft/sec, in one second you would be down to 32 ft/sec, and in 2 seconds down to zero. Sounds great, but if you were stopping your car at 1 g you certainly would feel it; everything that wasn't tied down would go flying forward, and if you had your seat belt on, you would likely have the breath knocked out of you. On the other hand, if you didn't have your seat belt on, things would be much worse.

About the maximum stopping rate a person in a passenger car can take is between .6 g and .8 g. Most of the time we stop with a force of about .2 g. What is the force on your body at these decelerations? Let's say you weigh 160 pounds. At .8 g you would feel a force of 128 pounds on your body (assuming you're strapped in); at .6 g it would be 96 pounds, and at .2 g it would be 32 pounds. Only the last would be a reasonably comfortable stop.

Of course, at .8 g you come to a stop much faster. It's easy to show that you will reduce your speed by 25.76 ft/sec each second, so if you are starting at 60 mph your speed in miles per hour will be 41.75 at 1 sec, 23.5 at 2 sec, and 5.25 at 3 sec. At .2 g these numbers become 55.6 at 1 sec, 51.2 at 2 sec, 46.8 at 3 sec, and 42.4 at 4 sec. All in all, it will take you more than 12 seconds to stop in this case, which may be longer than you wish.

Another thing we're interested in is stopping distance; in other words, the shortest distance we can stop without skidding or sliding. It is given by

$$s = v_0^2/2g\mu_s.$$

Note that we are dealing with two surfaces (the tire and the road) that are not sliding relative to each other, so we must use μ_s.

Again, substituting a few numbers, we find that for an initial speed of 60 mph (88 ft/sec) on dry concrete with $\mu_s = .8$ we get a stopping distance of 151.25 feet. How long would this take? We can determine this from the formula

$$s = \frac{1}{2} at^2.$$

We get 3.24 seconds. It's important to remember that this is the shortest stopping distance, and it's easy to show that it's approximately a .8 g stop, which would give a 126 pound force on your body (assuming you weighed 160 pounds). Needless to say, this would be quite uncomfortable.

Looking back at the coefficients of friction I gave earlier, we see that road condition is also important in stopping—in particular, whether the road is wet, dry, or icy. The condition of the tires is also important in this respect, as table 5 shows. (Note that the values in the table are approximate and depend on many factors.)

What does this tell us? Earlier we saw that the minimum stopping distance for a dry road at 60 mph was

Table 5. Coefficient of friction under various conditions for new and worn tires

Condition of Tires	Dry	Wet (light rain)	Heavy Rain (puddles)	Ice
		Weather (μ_s)		
At 60 mph				
New	.9	.60	.3	.050
Worn	.9	.20	.1	.005
At 80 mph				
New	.8	.55	.2	.005
Worn	.8	.20	.1	.001

151.25 feet. How much does this increase when we have water or ice? The results are as follows:

Light rain (new tires)	201.6 feet
Light rain (worn tires)	605 feet
Heavy rain (puddles and worn tires)	1210 feet
Ice (new or worn)	12,100 feet

These numbers should give you an appreciation for the dangers of wet and icy highways. I know from experience how hair-raising an icy road can be. When I was young I attended a college that was about 200 miles from my hometown, and part of the road went through a mountain pass that had a lot of snow on it during the winter. Each Christmas after my exams were over I would come home, usually with several other students in the car. I always worried about the road and had chains just in case I needed them. But I hated to put them on, because once I was over the pass I would have to stop and take them off.

We were traveling at about 60 on one trip when suddenly the back tires started to slide, perhaps because I had touched the brake. Once we started to skid, there was complete silence as I tried to regain control of the car. Nothing seemed to help, and I'm sure we were still doing close to 60 after several minutes had passed.

The car continued to move around, and soon we were going down the road backward, and it still seemed as if we had barely slowed down. The dropoff at the edge of the road was steep, to say the least, but at least there was a bank of snow between it and the road, so I wasn't too worried about it. But I *was* worried about meeting another car coming from the opposite direction.

Finally, after what seemed an eternity, the car came to rest. No one said anything but there were some heavy sighs. The car was now pointed in the wrong direction. We got the chains out and started putting them on, and as we did a large sanding truck approached. It stopped and the driver yelled, "You won't have to put your chains on now. It's sanded all the way into town." All at once everyone started to laugh. The driver must have thought we were a little crazy, until we told him we weren't going the way we were pointed. Needless to say, ice, with its coefficient of friction of .001 or less, can be a killer and we were very lucky.

Earlier I showed that the minimum distance you can stop a car, starting at 60 mph with $\mu_s = .8$, is about 151 feet. But if you look at the auto magazines you see numbers less than this. For example:

Model	Stopping Distance 60 to 0 (feet)
Ford Thunderbird	123
Audi A8L	124
Cadillac Seville STS	119
Chevy Camaro SS	120
Ford SVT Mustang Cobra	121

(continued)

(continued)

BMW 540i Sport	121
Lexus GS 430	115
Mercedes-Benz E430 Sport	116
Kia Reo	155
Acura TL	127
Saab 9³ Viggen	121
Volvo S60	119

These examples are obviously taken under ideal conditions. Why are they so much less than our figure? Brakes are, of course, continually improving and as a result the coefficient of friction of linings and tires is improving. These numbers are merely a measure of how high the coefficient is. We used $\mu_s = .8$ in our calculation, but higher values are now common in new cars. Of course, if you calculate the g force on your body for these stopping distances you'll find it is close to 1 g.

The Stopping Sequence

Several factors, aside from putting pressure on the brake pedal, are involved in stopping. A sudden stop usually begins with the realization that there is danger. Your foot shoots to the brake pedal and pushes on it. Several seconds later the car comes to a stop. If we look at the details, however, we see that there are several stages in the stopping sequence, and each takes a certain amount of time. Indeed, each of them may be short, but when your car is going 60 mph, it is moving a considerable distance in very short periods of time.

First of all, you have the shock time—the short period of time during which you realize there's danger. Your foot moves quickly toward the brake. How fast it gets there depends on your reaction time. Reaction times can vary considerably—from about .3 to 1.8 seconds; only

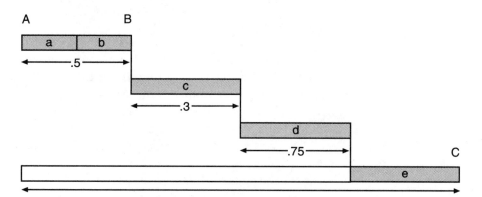

Fig. 40. The stopping sequence. A is the point at which danger is recognized; B is pedal application, or the start of braking, and C represents standstill; a is shock time, b is reaction time, c is application time, d is buildup time, and e is start of maximum braking effect; the arrow across the bottom shows the total time to stop.

highly skilled drivers have reaction times as low as .3 second. It also depends on factors such as youth, health, alcohol consumption, and distraction. Reaction time, along with shock time, can require a second or longer depending on your ability as a driver. Let's assume that you're a good driver and they take only .5 second. The next consideration is the application time, or the brake system's response delay, which can take another .3 second. Then there's the pressure buildup time, which can take .75 second. All these factors are shown in figure 40.

The Heat Machine

We don't usually think of a car as a heat machine, but that's what it is when it comes to slowing it down. A car in motion has kinetic energy, and, as we saw in chapter 2, the formula for kinetic energy is

$$KE = \frac{1}{2}mv^2.$$

Furthermore, we know that energy is conserved, so we can only change it in form. This means that if we want to stop our car we have to get rid of its kinetic energy, and the only way to do this is to convert it to some other form of energy. This "other" form is usually heat. When we press on the brake pedal we generate a lot of heat in the brake linings and in other parts of the brake.

Let's assume our car is going 60 mph. How much kinetic energy does it have? A simple calculation for a car of 2500 pounds gives:

$$KE = \tfrac{1}{2} mv^2 = \tfrac{1}{2} (W/g) \, v^2 = \tfrac{1}{2}(2500/32)$$
$$(88)^2 = 302,500 \text{ lb-ft.}$$

One BTU is 778 lb-ft, so we have a kinetic energy of 389 BTUs. This means we have to get rid of 389 BTUs of heat before the car stops. As the brakes are applied and heat is generated as explained above, they need to be cooled by air passing over them. The heat will, of course, also increase the temperature of the wheel. For average braking, its temperature will be about 350° F, but if you are forced to do a lot of braking it can increase to from 500° to 800° F.

Unless you're a truck driver up in the mountains you usually don't worry much about your brakes overheating, but there are times when this can be a problem. Once when we were in an RV coming down from a campground in the mountains, we went through many switchbacks. As we came down, I noticed my brakes acting up, and sure enough when I got out and felt them they were sizzling. I had to stop to rest the brakes many times before I got all the way down.

Brake Linings

Friction is, of course, important at the interface between the road and the tire, but it's also important at the brake pad or lining. The coefficients of friction of brake linings, or pads, vary considerably. Passenger cars usually have brake linings with coefficients from .25 to .35. If the coefficient of friction is less than .15 the braking power of the car will be poor, and if it is greater than .55 the brakes will tend to grab.

There are many factors that determine the quality of a brake lining. Some of them are fade resistance, wear on the rotor, the life of the lining, and quiet operation. Fade is a decrease of μ that can occur when the brakes get hot.

One of the most important things in a lining is the base frictional material: the small (or sometimes relatively large) fibers that provide friction and heat resistance. Most linings are metallic, semimetallic, or synthetic. Asbestos was used commonly for lining at one time, but it is no longer used. Most modern pads are high-tech and contain multifriction layers and carbon-metallic material. The newer materials stop better, resist brake fade, reduce brake pulsing, eliminate brake noise, reduce brake dust, and have a long pad/rotor life.

Tire Traction and Weight Transfer

Imagine you're coming into the last lap of the most important race of the year. You're in the lead but being pushed by the favorite, who is just a few feet behind, and you can see him in your peripheral vision out the side window. One more turn to the final stretch. You hit the accelerator, then brake, but you take the turn with a little too much speed. The crowd is on its feet. You feel your back tires starting to slide. A hollow sensation hits the pit of your stomach. You made a mistake. You struggle to regain control and as you do the favorite passes you. Seconds later you pass the finish line—in second place. You curse to yourself for allowing your tires to lose traction.

Fig. 41. Contact patch.

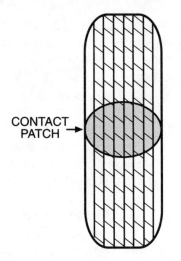

CONTACT
PATCH →

Traction is, indeed, important. It is a measure of how well the tire sticks to the road, and it depends on the *contact patch*—a roughly oval region of the tire that is in contact with the ground. Everything that happens to the tire and the car depends on this patch. Its position on the tire changes, of course, as the tire moves around; but it is always the same size and, assuming the tire is perfectly uniform, it is always the same. (See fig. 41.)

If you apply a force that is greater than the "adhesive" or holding force of the tire, it will begin to slide. We can determine when this happens from the formula $F = \mu W$. We know that $W = mg$, where g is the acceleration of gravity. All the weight of the car is not always on the tires; weight transfer can also be caused by braking, accelerating, or cornering. To take them into account we will apply a factor x. Thus, $W = xmg$. Now the acceleration due to the force F is $a = F/mx$ and therefore

$$a_{max} = F/mx = \mu W/mx = \mu xmg/mx, \text{ or } a_{max} = \mu g.$$

This is the maximum acceleration a tire can take before it slides, and surprisingly it is independent of the weight of the car. It is easy to show why explicitly. Using the above formulas, we can write

$$F/W = \mu = a_{max}/g = a_{max} \text{ (in terms of g).}$$

Now take a tire off your car and load it with some weight. Assume it is 100 pounds. Using a scale, pull on it until it moves and record the reading on the scale. This is F; W will be 100 pounds plus the weight of the tire. Taking their ratio, let's say you get $a_{max} = .9$ g. If you put 200 pounds on the tire and do the same thing, you will get a different F, but the ratio will still turn out to be .9 g. Thus, the acceleration before the tire slides is the same. In short, it will slide with the same acceleration, regardless of the weight involved. In practice, this is only approximately true, but for our purposes it is a good place to start.

With the above information we can define what is called a *traction* (or *friction*) *circle* (fig. 42). Let's assume we have a tire that slides at 1 g. We can draw a circle with acceleration in one direction, braking in the opposite direction, and cornering (left and right) perpendicular to them. We represent our motion by an arrow. From the circle in figure 42 we see that if the tire is using all its available traction to accelerate or brake, then no traction is available for cornering. If you try to corner, you will slide. Conversely, if you are cornering near your limit of 1 g, then there is no traction left for braking or accelerating, and if you try you will slip. As we will see in chapter 9, this traction circle is particularly important to race drivers.

Sliding can lead to a loss of control of the car and can therefore be dangerous. A small amount of controlled sliding, however, can be advantageous. A tire actually

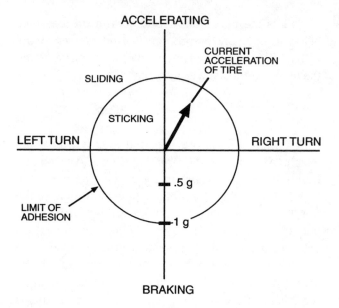

produces its highest traction when slipping by about 20% because the tread surface is moving into greater contact with the road surface. (A 20% slide gives only slight skid marks.) With sliding rates greater than 20%, however, the driver can lose control of the car.

Also important to drivers is the *slip angle,* which is the angular difference between the direction the wheel is traveling and the direction of the tread. Earlier I mentioned that the formula $F = \mu W$ is only approximately true because it applies only to nonelastic tires, and we know that tires are not nonelastic. In particular, they deform considerably under the forces to which they are subjected, and these deformations affect the slippage and create a slip angle.

Because a tire is elastic, it delivers force to the wheel and therefore to the car when it is stretched and twisted. There are several ways in which a tire can be deformed. First is *radial deformation,* which is visible as a bulge in the sidewall near the contact patch. Second is *axial deformation,* which tends to pull the tire off the rim. Third is *torsional deformation,* which is the difference in axial deformation from the back to the front of the contact patch. Finally, there is *circumferential deformation* around the tire. All of these have an effect on μ and therefore F and the slip angle. We will discuss slip angle in more detail later.

Also important in relation to sliding is the weight transfer referred to above. As we saw in chapter 2, acceleration transfers weight to the rear tires and leaves the front tires with less traction, while braking gives the front tires more traction and the rear tires less. Furthermore, cornering to the right gives the tire on the right more traction and the one on the left less traction, similarly, cornering to the left gives the left tire more traction and the right one less. When weight is transferred off a tire it is more likely to slide. For example, when the traction of the rear tires is reduced, a rear wheel slide can occur, with the rear end of the car sliding out in one direction. Such sliding is frequently associated with a locking up of the brakes.

In an earlier chapter we derived a formula for the amount of weight transfer. Using this formula we can calculate the weight transfer for a 3500-pound car with a 55/45% distribution of weight around the center of mass. For several different g's the distribution is as follows for front, back, and amount transferred, respectively: at .2 g, 59/41%, 132 pounds; at .4 g, 53/37%, 264 pounds; at .6 g, 67/33%, 397 pounds.

The Brake System and Hydraulics

So much for stopping, sliding, and traction. Let's consider the brake system itself. As we saw earlier, the first cars had mechanical brakes that were very inefficient and could be dangerous. The mid-1920s brought hydraulic brakes, a vast improvement over mechanical brakes. They decreased the stopping distance dramatically. The advantage of hydraulic brakes is that the force is the same on each of the brake pads, so the stopping power on the left side of the car is the same as on the right side, helping you to bring the car to an even and straight stop.

The overall brake system consists of the brake pad, disk brake unit, brake lines to the brakes, and the master cylinder. The brake lines run from the master cylinder to the wheel; they are usually made of steel, except near the wheels, where they are made of a strong flexible material because of the motion of the wheels. (See fig. 43.)

When you press the brake pedal, the master cylinder, which is basically a hydraulic pump, pressurizes the brake fluid and pushes it through the hollow steel tubes to the wheels. A valve reduces pressure to the front brakes initially so that the rear brakes begin to brake slightly before the front ones. The fluid in the system is under pressure, and this pressure is applied to small pistons in the calipers. This causes the brake pad to press against the rotor and the resulting friction stops the car.

The principle upon which hydraulic brakes is based was discovered in 1650 by the French physicist Blaise Pascal. It is referred to as Pascal's law:

The pressure exerted on a confined liquid is transmitted undiminished in all directions and acts with equal force on all equal areas.

The main reason hydraulic systems are used in cars is that fluids are incompressible and can easily flow through complicated paths. Pressure applied at one end

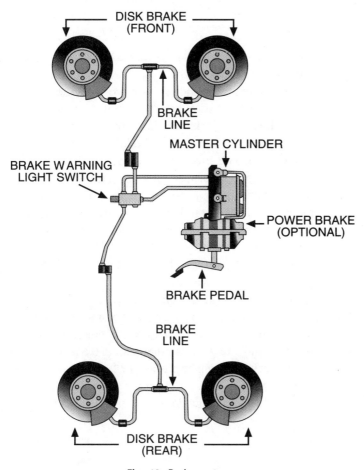

Fig. 43. Brake system.

of the system is therefore transmitted undiminished throughout the system. To understand this better, consider a small cylinder with a plunger in it, with fluid beyond the plunger. Assume the area of the base of the plunger is 4 square inches, and that we exert a force of 200 pounds on it. What is the pressure of the fluid? Pressure and force are related by the formula

$$p = F/a,$$

where p is pressure, F is force, and a is area. This means the pressure in the above case is 200/4 = 50 psi. The unit psi refers to pounds per square inch and is commonly used in hydraulics. Sometimes another unit, called the Pascal (or kiloPascal = 1000 Pascals), is used. They are related by 1 psi = 6.875 kP.

Pascal also discovered that liquids could be used to transmit motion. Because pressure is transmitted undiminished throughout the fluid, if we push down on a plunger at one point in a system for one inch, a plunger at a distant point with the same area will also be pushed one inch (fig. 44).

What happens if we change the size of the second plunger? Assume we have a plunger with a diameter of one inch. Its area is then $\pi r^2 = 3.141(.5)^2 = 0.785$ in^2, and if we push it one inch the volume is 0.783 in^3. In other words, 0.785 cubic inches of fluid is displaced. If the second plunger has a diameter of 0.5 inch, its area is 0.154 in^2 and it will be pushed 0.785/0.154 = 5 inches.

We can also change the force on a plunger in an enclosed system. Consider a system with three plungers

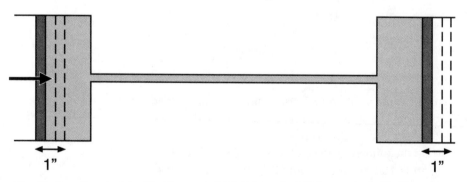

Fig. 44. Pascal's principle. If plunger on the left is pushed in by one inch the plunger on the right moves by the same amount, assuming it is the same size.

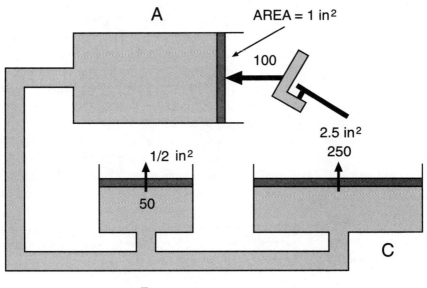

Fig. 45. Pascal's principle showing plungers of different sizes.

as shown in figure 45. Plunger A has an area of 1 in², B has an area of 0.5 in², and C has an area of 2.5 in². If we apply a 100-pound force to A we get, using $F = pa$, 50 pounds of force at B and 250 at C. Note, however, that the distance through which the force has been exerted changes in each case. The force at C is 250, but it is exerted over a distance only $\frac{1}{2.5}$ the distance the force was exerted.

Disk Brakes

Both disk brakes and drum brakes are used on cars. Some cars still have disk brakes in the front and drum brakes in the back, but drum brakes are being replaced more and more. Soon they will be a thing of the past, so I will say little about them.

In a disk brake the brake pads are squeezed against a metal rotor during braking. Basically, the brake consists of a flat metal rotor in the form of a disk that turns with the wheel, along with a stationary component that is mounted over it, called the *caliper.* Braking occurs when the caliper forces the pads against the two sides of the rotating disk. (See fig. 46.)

The caliper contains a piston or pistons along with brake pads. When the brakes are applied, the brake fluid flows to the brake and pushes on a piston, which in turn forces the pad against the disk or rotor. There are two types of calipers: fixed and floating. The fixed caliper remains fixed relative to the disk, and there are pistons on both sides of the disk. When the brakes are applied, both pads are pushed onto the rotor.

The second type of caliper, the floating caliper, is used much more commonly than the fixed variety; in fact, fixed caliper brakes are rarely used today. With the floating caliper there is usually only one piston, which

Fig. 46. Disk brakes.

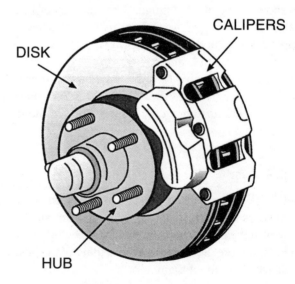

DISK

CALIPERS

HUB

is on one side of the rotor. When the brakes are applied, the pad is pushed via the piston into one side of the rotor. But at the same time, because the overall caliper can move, the pad on the other side is pulled *toward* the rotor.

There are both advantages and disadvantages to disk brakes. They are air cooled and stay much cooler than drum brakes, but normally power boosters are required with them.

ABS

ABS, or the antilock braking system, is designed to stop the skidding that occurs when brakes lock up. In most cases the system will allow you to bring your car to a safe stop. It was first used on cars in the late 1960s, and by the mid-1980s it had become common. It is now standard equipment on many new cars.

This braking system does not necessarily produce shorter stops on all surfaces. A stop with regular brakes on dry concrete will be about the same as an ABS stop. But when traction is lost on a wet or icy road, ABS produces much shorter stops, and, even more important, it usually allows you to keep control of your car.

ABS works by monitoring the speed of each wheel of the car. It determines which tires are turning properly and which are not—in other words, which tires have traction. If traction is lost on a particular tire, the ABS system will release the brakes on that wheel until traction has been regained.

Central to the ABS system is an electronic control unit (ECU); in a sense, it is the computer of the system. It receives signals from electronic sensors that are mounted in the wheels. If the rotation of the wheel suddenly decreases, the ECU reduces the hydraulic pressure to the brake to that wheel. It monitors each of the four wheels. The person applying the brake in the car will feel a pulsing of the pedal caused by the stopping and starting of the hydraulic pressure. Pulsing or pumping of this type can occur up to fifteen times per second.

It is important to keep a steady pressure on the brake pedal during the time the ABS system is activated. Under sliding or skidding conditions we are used to pumping the brakes, but this is no longer required. The ABS system does it for us.

The ABS system is a valuable addition to the modern car, but it can't do everything. Excess speed, sharp turning, and slamming on the brakes can still cause skids, so ABS can't prevent all skids. In fact, some skidding usually occurs before it is activated. It does, however, decrease the stopping distance considerably and helps you to keep control of your car.

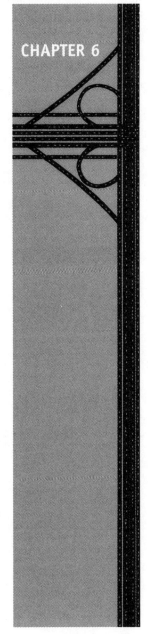

CHAPTER 6

Springs and Gears

The Suspension System and the Transmission

In the movie *Swordfish* John Travolta was chased by enemy agents in black Lincoln Navigators. He drove a sleek blue TVR Tuscan, a sportscar made by a relatively small car manufacturer in Bristol, England. In this exciting car chase, Travolta barely outmaneuvered the Navigators, one of which went flying through the air, flipping over and crashing into the roof of a diner. In the end Travolta's car was riddled with bullet holes, but naturally he escaped unharmed. With stunts like this, the cars would have to have good suspension systems. After all, it's the suspension system that cushions you from the jolting and bumping that are part of everyday driving. Even though this type of driving isn't part of most people's day, the suspension system is still an important part of the car. It connects the wheels of the car to the body, and in addition to making the ride smooth, it makes the vehicle safer to handle. In this chapter we deal with both the suspension system and the transmission. The transmission transfers power from the engine to the driveshaft and eventually to the wheels.

The Suspension System

Almost everyone has had a ride on a road filled with ruts and bumps that seemed to jar the daylights out of them. The road to our cabin is gravel and frequently develops "washboard," which sometimes forces me to slow almost to a stop to allow the vibrations to die down—and this with a modern suspension system. You can imagine how much worse it would be without such a system. Fortunately, most of the time we drive on smooth super-highways, or at least on roads with few ruts and irregularities in them. But almost any road has some bumps, and we would certainly feel them if it were not for our suspension system.

Suspension systems provide ride comfort but they also provide something that is just as important: they ensure that all four wheels of a vehicle are always in contact with the ground. If the wheels were attached to the chassis rigidly, at least one of them would not be in contact with the ground at any given time. And this, of course, would be dangerous.

Vibrations are an inevitable part of any ride, but not all vibrations are uncomfortable. Furthermore, ride comfort is affected by other things such as body roll, pitching of the car, and jerkiness. If the car rolls excessively when you go around a corner, or if it jerks back (squats) on acceleration, or pitches forward on braking (dive), the ride is quite uncomfortable. But vibrations in the car are usually considered to be the biggest annoyance, since they are continuous. All frequencies of vibrations, however, are not irritating. Most people agree that frequencies in the range 60 to 90 vibrations per minute (vpm) are comfortable. The reason, no doubt, is that this is close to the average person's walking rate, and therefore it's a body vibration that is familiar. Vibration rates such as 30 to 50 vpm, on the other hand, cause motion sickness in many people, and higher rates from, say, 200 up to 1200 vpm, are considered to be harsh and uncomfortable for almost everyone. The head and neck are particularly sensitive to rates from 1000 to 1200 vpm. High frequencies such as this are usually due to the tires or to axle vibrations.

The vibrational frequencies given above are for vertical vibrations, but there are also longitudinal vibrations in a car, and they can be uncomfortable as well. Strangely, some of the most uncomfortable longitudinal vibrations are in the range 60 to 120 vpm, which is the region of greatest comfort for vertical vibrations. Longitudinal vibrations are not as common as vertical vibrations, however, and occur only when the car pitches or rolls.

The vibrations of a car are associated with the springs in the car. Springs vary considerably in "springiness"—ranging from very soft to very hard. For a smooth, comfortable ride they need to be relatively soft, but if they are too soft the vehicle will undergo considerable "vertical travel." This is the distance the springs move as they go through their up-down cycle of compression and expansion. And a large amount of vertical travel makes the car hard to handle. A car with soft springs experiences considerable body roll when you take a corner too fast. In addition, it dives hard under heavy braking and squats when you accelerate fast. A compromise must therefore be made. The springs have to be hard enough so that the car is easy to handle, but soft enough so the ride is comfortable.

While the suspension system must provide comfort, it needs to do more than this. It must keep the wheels and tires upright so that maximum tire tread is on the road at all times. In addition, the car itself must remain as upright as possible at all times. This means that when you make a sharp turn, the suspension system should compensate for those forces on the car which tend to make it roll.

The job of the suspension system is complicated because the front and back wheels generally perform different functions. Except in a front-wheel-drive vehicle, the back wheels are associated with the driving of the vehicle and the front with the steering. Because of this arrangement, the front wheels have more complicated suspension needs.

One of the complications is called the *Ackermann effect,* named for Rudolph Ackermann, who took out a patent in England in the early 1800s for a system designed to compensate for the effect. (See fig. 47.) The two front wheels of a car are some distance apart, so that when a turn is made the inside front wheel turns at a sharper angle than the outside wheel. This is because the inside wheel moves around a circle of smaller radius than the outside wheel. The difference in angle

Fig. 47. The Ackermann effect. Note that the inner front wheel is at a greater angle than the outer one.

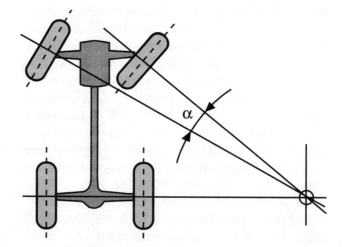

between the two wheels is called the *Ackermann angle,* and it becomes larger as the wheels are turned more sharply. At small-angled turns, the difference is usually of little significance, but at higher-angled turns it is important and must be compensated for. Several relatively simple types of steering correction mechanisms can compensate for it. In most cases, they do not compensate exactly for all turning angles, but the tires and free play in the steering components make up for the rest. As we saw earlier, tires deform considerably when the car is turned, and rubber mountings are used in many of the connections within the suspension system, and they help.

Linking Up: Attaching the Wheels to the Car

We refer to the body of the automobile as the sprung system. The wheels, suspension linkages, and other components of the suspension system are the unsprung system. In essence, the unsprung system is on the road side of the springs; the sprung system is on the other side. This means that the unsprung weight is not damped and therefore reacts directly to road irregularities. The tire, of course, provides some cushioning, but it is a relatively small amount. The sprung system, on the other hand, is filtered from road disturbances by the suspension system.

For maximum comfort and best handling, the ratio of unsprung mass to sprung mass should be as low as possible. In most cases, unsprung mass is 13% to 15% of the total vehicle weight. This means a 3000-pound car has an unsprung weight of about 450 pounds, and this 450 pounds is reacting directly to the road irregularities. The forces generated by it can have a detrimental effect on the ride and handling of the car.

One way to keep the ratio low is to use an independent suspension system—in other words, a system that allows the wheels to move independently. This means that when one wheel hits a bump, the other wheels are not affected. Wheel independence is also important in keeping the four wheels on the ground at all times. All vehicles now have independent front suspension systems and a few have independent systems on all four wheels.

Let's start with a brief overview of the suspension system, beginning with how the suspension system is connected to the wheels. The connectors are called *control arms,* and there are several ways the connections can be made. One of the most common is the *double wishbone,* or *double A-arm.* In this system the two arms are in the shape of the letter "A," with one mounted above the other (fig. 48). In the early systems of this type, the A-arms were parallel and of equal length. This caused the wheels to lean out in turns, which resulted in excessive tire scrubbing (scuffing). It

Fig. 48. Double A-arm, or wishbone suspension.

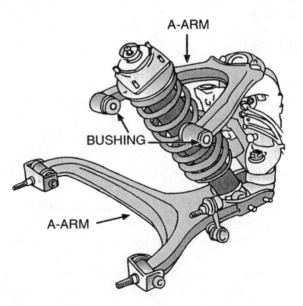

didn't take long, however, to discover that nonparallel and unequal-length arms were much better. Indeed, with unequal arm lengths, engineers were able to design systems that provided almost infinite control over the movements of the wheels. The upper arm is invariably the shorter of the two. The tips of the arms hold ball joints that allow the wheel to rotate and pivot. Rubber bushings are used on the other ends, and coil springs are usually mounted between the A-arms.

Another type of system that is commonly used is the MacPherson suspension (fig. 49). It was invented in 1945 by Earl MacPherson of the Ford Motor Company and was first used on the 1950 English Ford. In this system, the upper A-arm of the double wishbone system is replaced with a tall shock-absorbing strut attached to the body structure. A coil spring is usually mounted over the shock absorber assembly. The strut stands on a base that extends from the wheel hub. The MacPherson system is relatively simple and has therefore

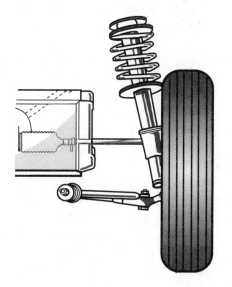

Fig. 49. MacPherson suspension.

become quite popular. It has a disadvantage, however, in that the strut is relatively long and requires a high hood and fender.

Finally, we have the multilink system, which is being used more and more in modern cars. In this system, a number of links or arms give support to the wheels. Because there are several links, this suspension can be highly tuned to give excellent handling characteristics.

Computerized Suspension Systems

In conventional suspension systems the characteristics of the ride and handling are set by the designer, and they are, for the most part, fixed. There are systems, however, in which these characteristics can be changed, either manually by the driver or automatically by a computer system. These electronic systems, as they are called, center around a computer. Sensors are set up throughout the suspension system and information is fed to the computer. Things such as roll, pitch, height of car above the road, speed of turn, turn radius, and angular velocity of the wheels are monitored. In the more advanced systems, the computer checks over the data and automatically changes the suspension system according to the conditions. In simpler systems, a change between a soft ride and a hard one with better handling is usually all that can be made, and it is made manually.

One of the more advanced of these systems is currently being used by Mercedes-Benz; it is referred to as Active Body Control, or ABC. In this system hydraulic adjustable servo cylinders in the struts of the suspension system counteract the forces that would normally be transferred to the body. The system can give a soft or hard ride, or almost anything in between; several adjustments for pitch and roll are also possible. Thirteen sensors are used to send information to the computer.

With the success of this system, most other auto manu-facturers are looking into similar systems, and it's only a matter of time before they become common in cars.

A Little "Roll and Rock"

When a car goes around a corner it experiences a cen-tripetal force that tends to make it roll. This is the force you feel on your hand when you whirl a ball on a string. It is countered by a frictional force between the tires and the road. When a car experiences a centripetal force it leans, and it's up to the suspension system to keep this lean to a minimum.

Lean is usually referred to as roll, and it takes place around an axis called the *roll axis*. For most cars it's rel-atively easy to determine the position of this axis. The first step is to determine the *roll center* of the front wheels (fig. 50). The procedure is different for the two types of systems, so I'll begin with the double A-arm system. In this system you have two arms that can be slanted at various angles. We'll assume they are slanted

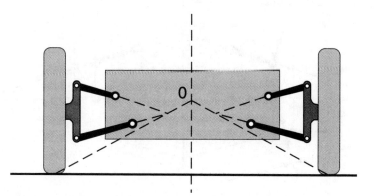

Fig. 50. Determining the roll center. In this case it is at 0, which is above ground.

as shown in figure 50. This is only one possibility of several, but it will illustrate the technique. If you extend the two arms using dotted lines as in the figure, they will cross at a point called the *instantaneous center.* It is so called because it changes when the suspension system is compressed or raised, so it is not the same at all times. For our purposes, however, this won't pose a problem. Each wheel has one of these centers. Now draw a line from the bottom of the tire through the instantaneous center, and do the same thing for the other side. The two lines will cross at the roll center. This is the point around which roll takes place for these wheels.

In practice, the roll center can end up anywhere. In most cases it's above ground, as in the above case, but it can be at ground level—this would be the case if the A-arms were parallel—and it can be underground (fig. 51).

With the MacPherson system we don't have an upper A-arm so we have to determine the roll center differently. In this case we use the angle of the lower A-arm as before, but we now use a line that runs perpendicular to the spring axis as the second line. Again we

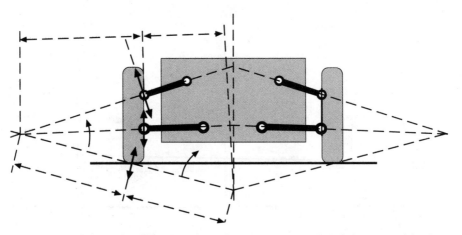

Fig. 51. Roll center for different angles of the A-arms. Here it is below ground level.

extend the two lines until they cross and draw a line from the bottom of the tire to this point.

Both the front and the back wheels have roll centers and in general they will be at different heights. In most cases the back roll center will be higher. To determine the roll axis, which is what we originally were interested in, we draw a line between the front and rear roll axes, as shown in figure 52. This is the axis that the car is trying to roll around when the car rounds a corner.

It's easy to see why the car would have a force on it as it corners, but why would it necessarily roll? For roll you would need a torque. Let's take a closer look. Earlier we talked about the center of gravity of the car. This is the point where all of the weight of the car is acting, and in general it is not at the same point as the roll axis. Since the centripetal force acts at the center of gravity and the twisting force is around the roll axis, we do, indeed, have a torque. Note that the lever arm of the torque is the perpendicular distance between the two points. The torque is called the *roll couple,* or *roll torque* (fig. 53).

In summary, then, the roll couple causes roll around the roll center, and the springs take up the resulting torque and are compressed. There are therefore two forces—one causing the lean and one countering it.

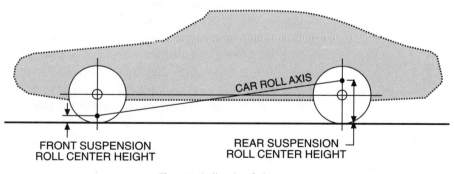

Fig. 52. Roll axis of the car.

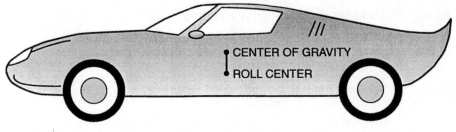

Fig. 53. There is a roll torque between the roll center and the center of gravity.

Because they are not exactly equal there is some lean, but it is usually small. The torque causing the lean is called the *resultant.*

As we just saw, the roll axis can be slanted downward toward the front of the car or downward toward the back of the car, and the handling characteristics of the car will depend on which of these is the case. If it is slanted toward the front of the car, we will have considerable traction on the front wheels of the car and less on the back. The result will be oversteering. If it is slanted toward the back of the car we will have considerable traction on the back wheels and less on the front. The result in this case will be understeering.

Another important component of the suspension system is the stabilizers. I will discuss only one type, namely the antiroll bar. It connects either the two front wheels or the two rear wheels (or both), and its job is to stop the vehicle from leaning too much during turns. It doesn't affect the suspension system itself, and only has an effect when the vertical movement on one side of the vehicle exceeds that on the other. When this happens, the antiroll bar exerts an opposing force, which restricts the lean. The bar also reduces the cornering force and adhesion of the tires.

Finally, we have the pivot axes, which are located on the control arms. When the brakes are suddenly applied, the car tends to dive (the front end is lowered), and when it is accelerated suddenly it squats (the back end is lowered). The pivot axes have antidive and antisquat configurations in them to control these tendencies.

In relation to roll, a design engineer is concerned with what is called *roll stiffness,* or the torque exerted by the suspension system as it tries to pull the body back to its normal upright position. For a given side force, the roll stiffness depends on the height of the roll centers, the spring stiffness, and the effects of the antiroll bar. By making changes to any of these the design engineer can change the roll stiffness of the vehicle.

Suspension systems vary in modern cars. Table 6 shows what is used in several representative vehicles.

Table 6. Suspension systems for several 2002 cars

Model	Type of Suspension (front/rear)
Ford Focus ZTS	Control arms (A), coil springs, dampers, antiroll bar / multilink, coil springs, damper, antiroll bar
Audi A8L	Multilink, coil springs, antiroll bar / multilink, coil springs, antiroll bar
Cadillac Seville STS	Electronically controlled MacPherson struts, coil springs, antiroll bar / multilink, coil springs, electronically controlled shocks, antiroll bar
BMW 540i Sport	MacPherson struts, coil springs, antiroll bar / multilink, coil springs, antiroll bar
Lexus GS 430	Upper and lower control arms (A), coil springs, antiroll bar / upper and lower control arms, coil springs, antiroll bar
Dodge Stratus ES	Upper and lower control arms, coil springs, antiroll bar / multilink, coil springs, antiroll bar
Honda Accord EX	Upper and lower control arms, coil springs, antiroll bar / multilink, coil springs, antiroll bar

Springs for Less Bounce

The springs are a key component of any suspension system. Both leaf and coil springs are used in suspension systems, but we will take a look at only coil springs. Associated with any spring is a spring constant, which is usually designated as k. It gives a measure of the stiffness of the spring. High k designates stiff springs, low k, soft springs. The relationship is

$$F = kx,$$

where F is the force exerted on the spring and x is the distance it is extended or compressed (fig. 54). As an

Fig. 54. Weight attached to a spring. The spring is extended a distance x.

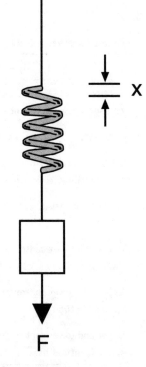

example, assume a 200-pound man compresses the spring of a car by 2 inches when he sits in it. The spring coefficient k is

$$k = F/x = 200/2 = 100 \text{ pounds/inch.}$$

When the tires of a car hit a bump, a force is exerted on the spring and it vibrates. In effect, the spring oscillates or vibrates back and forth with a certain frequency, and we can easily calculate this frequency. The period T, or time for one complete vibration or cycle of the spring, is given by

$$T = 2\pi \ (m/k)^{\frac{1}{2}}.$$

To illustrate, assume again that our 200-pound man compresses the spring in a car as above, and the car weighs 2500 pounds. We determined that k was 100 pounds/inch, therefore

$$T = 2\pi \ (2700/3200)^{\frac{1}{2}} = 5.77 \text{ seconds.}$$

The frequency of vibration v is $1/T$, and this gives us 10 vpm.

In practice, it is not just the springs that determine the frequencies of vibration you feel within a moving car. The springs are connected to various suspension linkages, so we have to deal with the overall system. The rate at which the overall suspension system compresses in response to weight is referred to as the static deflection rate, and it is this rate that determines the natural frequency of the vehicle. The design engineer must know what this frequency is, for it's a frequency to be avoided. If the frequency that occurs as the wheels pass over bumps and irregularities in the road coincides with the resonant frequency, the suspension

system not only won't absorb the bumps, it will magnify them.

The natural frequency of a suspension system can be obtained from

$$f_n = 188/d_s,$$

where d_s is the static deflection in inches and f_n is in vpm.

Regardless of whether the bump is at or near resonance, once the suspension is compressed it will continue to oscillate, and if this were allowed to continue the ride would be very uncomfortable. These oscillations obviously have to be damped out as quickly as possible, and this is the job of the shock absorbers.

Absorbing the Shock

When shock absorbers, which are located near each wheel, damp the springs, the oscillations last for a cycle or less. There are several types of shock absorbers, but in general they consist of a piston fitted tightly inside an oil-filled cylinder. As the car encounters ruts and bumps in the road, the piston moves up and down in the cylinder, and since the cylinder is attached to the car body, the car remains relatively stationary. When the piston moves inside the cylinder, valves let oil pass from one side of the piston to the other. The piston has small holes or valves in it. As a result, the speed with which it passes from one side to the other is regulated. The speed with which the piston moves depends on the size of the valves. Large valves allow it to move faster, small ones slow it down.

The Transmission

With all its gears and gear combinations, the transmission is one of the most complicated parts of the car. It's

also a major part of the power train that channels power to the rear wheels. There are two types of transmissions: manual and automatic. Nowadays most cars have automatic transmissions and most of the discussion of this section will therefore be about them. In manual transmissions, gear selection is made manually with a gear shift, and coupling and engaging are done using a clutch. In the automatic transmission, on the other hand, everything is done automatically, and the clutch is replaced by a torque converter.

The transmission is basically a torque-transferring and multiplying device that has a reverse gear and can be used for braking. It is usually located directly behind the engine (in some cases it is located farther back). Braking by means of the transmission is important for large trucks, particularly when they are going downhill. No doubt you have passed a few as they were gearing down on steep hills.

The engine is the major torque-generating device in the vehicle. The pistons rotate the crankshaft, generating torque that is supplied to the back wheels. This torque cannot be passed directly to the back wheels because of the many different conditions that exist in a moving car. For example, if the load on the engine becomes too great for it to handle, the car will stall. The transmission overcomes this problem by increasing the torque.

One reason engine torque is not sufficient to propel the vehicle under all driving conditions is because of engine operating characteristics. A plot of torque versus rpms for a typical engine is shown in figure 55. We see that maximum torque is not reached at maximum rpms as you might expect; rather, it comes at about 50% to 60% of the engine's maximum, and this has an effect on its performance.

The torque output of the engine depends on several factors. First of all, it depends on whether the engine is

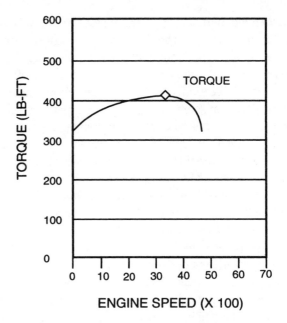

Fig. 55. Torque versus engine speed.

operating under a load. For similar conditions, the torque output under a load will be higher. Second, the engine will only provide the torque that it needs at any time. Finally, as we saw earlier, if the load becomes too great the engine will stall.

The maximum torque of an engine depends on several factors, such as size of cylinder, compression ratio, piston stroke, and richness of air-fuel mixture. When you press down the accelerator a certain air-fuel mixture enters the cylinder creating a particular torque. As you continue pressing it down, the richness of the air-fuel mixture increases and the output torque increases until it reaches its maximum. If the load becomes too great for this torque an increase is needed, and this is where the transmission comes in. It shifts to a lower gear, and this, in turn increases the torque. In the automatic transmission this is done automatically.

Let's take a brief look at the main features of the transmission. The transmission gives several different ranges that are provided by gear ratios between the various gears. The automatic transmission selects the proper gear ratio according to the engine load, speed, and the vehicle speed. Within the transmission are the planetary gears, which are unique devices for obtaining many different gear ratios and therefore several different forward gears and a reverse. If the gear ratio is, say, 3:1, this means that the engine rotates three times to every turn of the output shaft. We are mainly interested in the overall ratio between the input and output of the transmission.

The transmission is a torque multiplier, but it is not the only one. Torque conversion is also possible in the torque converter; in this case the conversion is continuous, however, in that any ratio is possible. Finally, the ring and gear pinions associated with the back wheels also give a fixed gear ratio—usually between 3:1 and 5:1. For the overall gear ratio we have to multiply all of these numbers together.

Gears and How They Work

The gears within the transmission are circular disks with teeth cut into them. They have to be made of high-quality steel to withstand the force impressed on them. In the simplest gears the teeth are cut straight across, but all the gears now used in transmissions have teeth that are cut at an angle. This increases the gear's strength and gives quieter operation.

When two gears of different radius mesh with one another, we have what is called a mechanical advantage (usually designated as MA). It is defined as

$$MA = (\text{output force})/(\text{input force}).$$

The best way to understand it is to consider a lever and fulcrum. If we apply a force F_1 to one end of the lever (as shown in fig. 56), considerable force (F_2) is exerted at the other end. We would use something like this, for example, if we wanted to move a large rock that we couldn't lift. Torque, as we saw earlier, is defined mathematically as Fl, and in a system such as that above we have

$$F_1 l_1 = F_2 l_2.$$

Let's assume that l_1 is seven times l_2 and that the force we are applying, namely F_1, is 50 pounds. We will get a lifting force of

$$F_2 = (l_1/l_2)F_1 = 50 \ (7) = 350 \text{ pounds.}$$

Let's apply this to a set of two gears. Assume the radius of the larger gear is r_2 and that the smaller gear is r_1. As an example, take r_1 to be 2 inches and r_2 to be 6 inches. Assume we apply a torque of 100 lb-ft to the

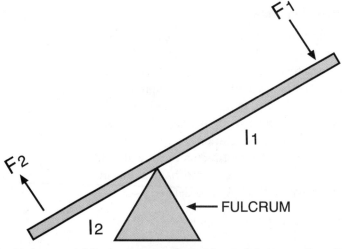

Fig. 56. Lever and fulcrum. F_1 is the applied force; F_2 is the resultant force; l_1 and l_2 are lengths.

smaller gear. We can easily calculate the force that is applied to the larger gear at the end of the gear tooth. Since torque is force times perpendicular distance, we have

$$T_1 = 100 = F_1 r_1 = F_1(2/12),$$

or $F_1 = 600$ pounds, where T is torque (fig. 57). This force is applied to the end of the tooth of the second gear, which is 6 inches from its center, or fulcrum point. Thus the torque of the larger gear is

$$T_2 = F_1 r_2 = 600 \ (6/12) = 300 \ \text{lb-ft.}$$

The input torque is therefore 100 lb-ft and the output torque (the torque of the larger gear) is 300 lb-ft. This means the torque associated with the larger gear is three times as great. The MA (torque) is therefore 3.

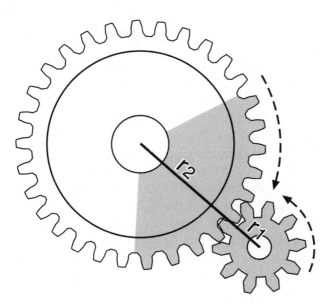

Fig. 57. Two gears meshing, one of radius r_1 and the other of radius r_2.

An easier way to determine the MA is to count the teeth in the two gears. If the larger one has N teeth and the smaller one has n, the MA is N/n. For example, in the above case if the number of teeth on the smaller gear were 8 and on the larger gear 24 we could determine the output torque immediately from the ratio 24/8.

It's important to note that in the case of a torque increase there will always be a speed reduction in the gears. If, for example, the torque output of the larger gear is three times that of the smaller gear, the smaller gear will be going around three times as fast as the larger one. In short, its rpms will be three times as great.

Planetary Gears

As we saw, the major component of the transmission is the planetary gear set. It is an ingenious combination of gears that gives many different gear ratios. It provides three or more forward gears and a reverse. The system is named for the solar system because of its resemblance to it. The central gear is the sun gear, and around it are three planetary gears. The planetary gears mesh with a ring or annular gear that is on the outside. (See fig. 58.)

The gears remain in contact with one another at all times, so there isn't the grinding of gears that frequently occurs with manual shifting. Each of the gears is on its own pivot and the three planetary gears are supported by a carrier. The planetary gears mesh with the ring or annular gear that has teeth on its inner circumference.

The planetary gear set can produce six different gear ratios along with a direct drive. The various gear ratios are obtained by holding one or another of the components of the system stationary, and making one an input and the other an output. For direct drive, which is the highest gear, the input and output must have the same speed.

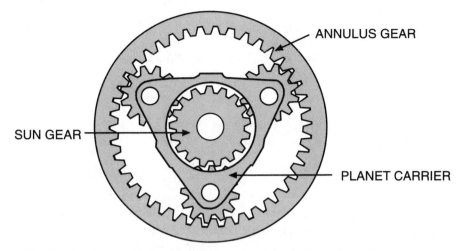

ANNULUS GEAR

SUN GEAR

PLANET CARRIER

Fig. 58. Planetary gears showing sun gear, planetary carrier, and annulus or ring gear.

I won't calculate the ratio in all cases but will illustrate with a couple of examples. You should be able to calculate the rest from them. As a first example, assume the input shaft drives the ring gear, and the sun gear drives the output shaft. In this case the carrier is stationary. To obtain the gear ratio you have to divide the number of teeth on the output gear by the number on the input gear. Let's assume the ring gear has forty teeth and the sun gear has eighteen. We will then have

$$\frac{\text{output gear (sun)}}{\text{input gear (ring)}} \quad \frac{18}{40} = 0.45,$$

and the ratio is .45:1. This means the output torque is less than the input torque, so the speed of the output shaft will be greater.

As a second example, assume the input shaft drives the planetary carrier and the ring gear drives the output shaft. Again, for the gear ratio we take the number of teeth on the output gear divided by the number on the input gear, but in this case the input is the planetary carrier. It doesn't have teeth. The number associated with it is the sum of the number of teeth of the ring and sun gears. We therefore have

$$\frac{\text{output gear (ring)}}{\text{input gear (ring + sun)}} \quad \frac{40}{40 + 18} = .69.$$

The ratio is therefore .69:1. Similarly, if the input shaft drives the ring gear and the carrier drives the output shaft, we get a ratio of 1.45:1. In this case there is a torque increase and a reduction in the output shaft speed. The other gear ratios can all be calculated in the same way. In all there are six possible combinations, each with a different gear ratio.

The gears, of course, have to be held in place, and this is done using multiple-disk clutches, one-way clutches, and friction bands. They hold one member of the gear set fixed, using hydraulic pressure. The pressure is routed to the correct holding element by a valve body that contains several hydraulic valves and contributes the pressure of the fluid. In general, they are referred to as *reaction members* of the transmission.

Compound Planetary Gears

In practice, manufacturers usually fasten one of the three members to the output shaft, limiting the number of ratios to two forward gears, or one forward gear and a reverse. Of course, this isn't enough. To overcome this lack, planetary gears are compounded; in other words, two planetary gear systems are used in conjunction with each other. There are two common designs for compound gear sets: the Simpson gear set and the Ravigeaux.

In the Simpson gear set, two planetary gear sets share a common sun. The Ravigeaux gear set has two sun gears, two sets of planetary gears, and a common ring gear. With either of these compound gear sets, three forward gears plus a reverse can be obtained.

Converting the Torque

The torque converter transfers torque from the engine to the transmission. It takes the place of the clutch in a manual transmission and operates through hydraulic forces that are provided by a rotating transmission fluid. It automatically couples and uncouples power from the engine to the transmission according to the engine rpms. When the engine is idling, the fluid flow around the device is insufficient for power to be transferred, but when the rpms increase, the fluid flow increases, creating a hydraulic force between vanes that allows power to flow. It is transmitted from the engine to the input shaft of the transmission, where it couples to the planetary gears. (See fig. 59.)

The main components of the torque converter are the

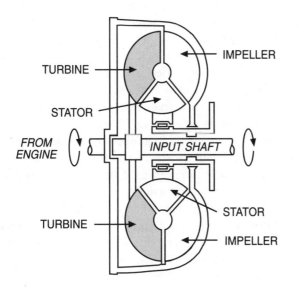

Fig. 59. Cross-sectional view of the torque converter showing impeller, stator, and turbine.

impeller, turbine, stator, and cover. The cover is connected to the flywheel and transmits power from the engine to the torque converter. The impeller is driven by the converter cover, and it, in turn, drives the turbine. The turbine is connected to the input shaft, and some torque multiplication can occur here. The stator aids in reducing the fluid flow from the impeller to the turbine and is equipped with a one-way clutch that allows it to remain stationary during maximum torque development.

The impeller drives the turbine just as one fan, which is plugged in and running, would drive a second fan that is not plugged in. The impeller and turbine both have blades and move in the same direction.

The torque converter provides a gradual, smooth connection between the engine and the load.

Continuously Variable Transmissions

Why do gear ratios have to be discrete? Actually, they don't have to be. There is a system that will give any gear ratio. It's referred to as the continuously variable transmission (CVT). Invented in 1958 by Hubertus van Doome, it has had problems, but within the last few years more and more car manufacturers are looking at it seriously. It was first used in the Subaru in 1989, and again in the 1996 Honda Civic HX.

Unlike traditional transmissions, the CVT does not use gears. Rather, a pair of variable diameter steel pulleys are connected by a metal belt. A computer within the power train selects the required gear ratio according to the vehicle requirements. One of the benefits of the system is smoothness during acceleration. Furthermore, it has been shown to be fuel efficient.

Almost all of the major car manufacturers now have CVTs under development. It is scheduled for use in the Audi A4 and A6, Honda Civic GX and HX, Honda Insight, and Saturn Vue.

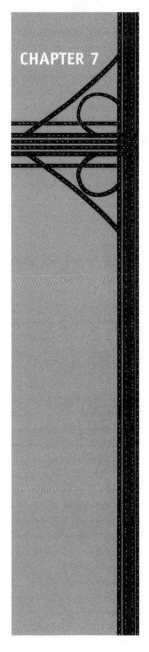

CHAPTER 7

What a Drag

Aerodynamic Design

Wow! Your eyes pop as you enter the main room of the latest auto show. You are particularly anxious to see the new Thunderbird. And sure enough—there it is. The lines . . . so streamlined. The front is like a flattened teardrop with the rest of the car tapering back to low, fabulous fenders. The enormous wrap-around wind-shield is like nothing you've ever seen before. It is, indeed, a sleek, beautifully designed aerodynamic car.

As you look around at the other cars on the floor, you think that they are much more streamlined than the cars of only a few years ago. The cars of today are, indeed, much more aerodynamic than those of even a few years ago, and one of the main reasons is that aero-dynamic cars consume less fuel.

What is aerodynamics? To the physicist it's the branch of physics that deals with the interaction between an object and the air around it as it passes through this air. A thorough understanding of it is crit-ical to airplane flight, and in recent years it has become increasingly important in relation to cars. The forces created by the airflow around a car depend on several factors: the shape of the car, the relative velocity of the air and the car, and other things such as protuberances on the car. At low speeds these forces are usually low, but at high speeds they can severely affect the performance of the car. Stability, tire traction, handling ability are all affected, and of particular importance, fuel economy is also affected. Engine power has to overcome aerodynamic forces, and this takes fuel.

Aerodynamicists are mainly interested in the *aerodynamic drag* on the car. The overall aerodynamic drag consists of five different types of drag: form, lift, surface friction, interference, and internal flow. I discuss each of these in detail later; for now, a brief description will suffice. *Form drag* depends on the form or shape of the car: how smoothly air passes over the contour of the car and how it breaks away at the rear. *Lift drag* is the result of pressure differences on the bottom and top of the car that create lift. *Surface friction drag* is a result of the viscosity of the air—in other words, how much friction there is between the various layers of air near the car. *Interference drag* is caused by projections on the car body, and *internal drag* is caused by air passing through the car.

The amount of drag caused by the various forces varies considerably. Percentage-wise, for an average passenger car we can assume the following drags: form, 55%; interference, 16%; internal, 12%; surface friction, 10%; lift, 7%.

Because early cars had relatively low speed, there was little interest in their aerodynamics. The first interest came in Germany, giving that country a ten-year lead on other nations. Germany built several large wind tunnels for use in testing their fighter planes in World War I. The treaty at the end of the war banned the design and testing of new planes, so they decided to use the wind tunnels to test the aerodynamics of cars. As a result they made several important breakthroughs. In 1921, for example, Edmund Rumpler introduced his Tropfenwagen, which had an aerodynamic drag less than a third that of most cars at the time. It wasn't, however, the most comfortable vehicle to ride in. A couple of years later, in 1923, Paul Jerey, a chief designer at the Zeppelin airship works, introduced the J-shape (the J stands for Jerey, not the shape of the car). (See fig. 60.) He showed that a car with a gradually sloping back had a drag about half that of other cars. But again there were problems with passenger comfort. His design, however, was the basis for several of the fastback models of the 1940s and 1950s, including the Citroën and Porsche.

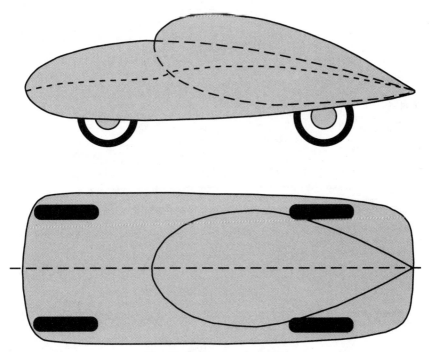

Fig. 60. J-shape, or fastback.

In late 1935 another important breakthrough was made at the Aerotech Institute in Germany. An engineer was testing a vehicle with a long, rounded tail. Disappointed with the results he was getting, he cut the tail off in an act of desperation. When he retested it he found to his surprise that the drag had improved; in particular, the tail didn't seem to affect the result. About the same time, Dr. Wunibald Kamm of Stuttgart showed both theoretically and practically that a blunt rear end decreased the overall drag. We now refer to this as the K-shape, after Kamm. (See fig. 61.)

Work in Germany was followed by similar work in France and Italy in the early 1920s. But it was not until the late 1920s that the United States got into the act.

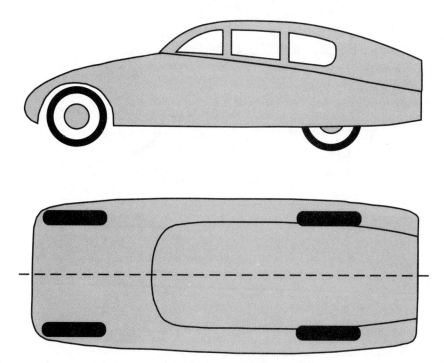

Fig. 61. K-shape, referred to as bobtailing.

Some of the earliest work was done in Michigan at the University of Detroit, and one of the most interesting results obtained at that time was by W. E. Lay. He showed that the shape of the tail end of a car was of little significance if the front end wasn't properly shaped. He measured the drag of a square box on wheels, then rounded the corners, streamlining them, and found to his surprise that the drag dropped to half its previous value.

Chrysler began taking an interest in automobile aerodynamics in the 1930s, and Ford a little later. But little serious work was done on passenger cars. The major interest at the time was racing cars. Stock-car driver Smokey Yunick was one of the first to consider aerodynamics. He realized that the underside of his car was creating considerable drag and attached a belly pan across it. It was so successful that NASCAR officials promptly banned it. Yunick continued to look for other advantages, and soon other drivers were following his lead. Chrysler began testing its race cars in wind tunnels, developing the Charger 500 and Dodge Daytona as a result. They were so successful that others quickly followed.

Passenger Car Aerodynamics

Even though sporadic testing was done on passenger cars, none of the large car manufacturers worried much about aerodynamics until the 1960s. Gas was cheap. Big fins, large fenders, and high blunt hoods were all the rage, and because they sold cars, no one wanted to upset the applecart. But wind tunnels soon showed that large fenders and fins were far from aerodynamic, and car manufacturers realized that fuel economy was being reduced. But it took the gas crunch of the early 1970s to force them into action.

How do you go about testing the aerodynamics of a car? Everything, it turns out, hinges on a simple number called the *coefficient of drag* (c_d). If you know this number, you know most of what there is to be known about the aerodynamics of the car. We'll leave the details of how it is calculated until later, but we'll look into what it tells us about cars.

The coefficient of drag is based on the drag of a flat square sheet; in the wind it has a c_d of 1.00. It was assumed early on that this was the maximum, but later it was shown that other shapes actually produced larger c_d's. Early cars typically had a c_d of .7. Over the years it has gradually decreased until some of the lowest values today are less than .3. As a rough guide we can say that a car has poor aerodynamics if it has a c_d of .5, moderate aerodynamics (but generally unacceptable today) if its c_d is .4 and good aerodynamics if its c_d is .3 or less.

In the mid-1980s, c_d's ranged from .5 to .3. Strangely, some of the largest luxury cars had the highest c_d's. The Chrysler Fifth Avenue had a c_d of .48. The Audi 500S, on the other hand, had a c_d of .33. The lowest c_d's did not decrease much through the 1990s, and the best cars still have a c_d of around .3. The main difference is that more cars now have c_d's in this range. This is surprising given that Rumpler's Tropfenwagen of 1921 had a c_d of .28.

With fuel economy becoming of increasing concern, aerodynamic styling is taking center stage. It has been pointed out, for example, that a 10% increase in c_d improves gas mileage by 5%. This means that an improvement from .4 to .3 in most cars would reduce gas consumption in the United States by 10%, saving tens of billions of gallons.

What are some of the c_d's of modern cars? They aren't reported as much as horsepower, torque, or time from 0 to 60, but manufacturers do occasionally publish them. Here are a few: Jaguar X-type 2.5 Sp, .33; 2000 Mazda MBV ES, .34; 2000 BMW M3, .33; 2002 Mercedes-

Benz C32 AMG, .27; Audi S4 amt, .32; 2002 Infiniti Q45, .30; Honda Insight, .25.

These are, of course, all streamlined passenger or sports cars. Many of the vehicles on the road are SUVs and trucks, and their c_d's are much higher. In most cases, they are not published.

Streamlines and Airflow over the Car

A small section of the airflow over a car is referred to as a *streamline,* and the family of streamlines is called the *airflow pattern.* This pattern depends on the car's shape and the speed of the car through the air. It can be seen in a wind tunnel if an opaque gas such as smoke is used.

Streamlines in the vicinity of a car are complex. Over the front part of a car they generally follow the contours of the car, but they can split and separate. Of particular importance is the internal friction, or viscosity, of the air. It was shown in 1744 by Jean LeRond D'Alembert of France that if the viscosity is zero, no tangential forces can act on the surface of an object, and no forces would therefore be exchanged between the air and the object; in other words, aerodynamic forces would not exist. This strange result is known as *D'Alembert's paradox.* Of course, in practice, no fluids have zero viscosity, so aerodynamic forces do exist.

Indeed, viscosity creates forces between the layers of air that pass over one another. These forces are frictional forces, and they give rise to what is called the *boundary layer* (fig. 62). The layer of air in contact with the surface of the car tends to stick to it so that it moves along with it. The next layer gets dragged along because of the friction, but it lags behind. The third layer lags even farther behind, and so on. Finally, as we continue outward, the air is at rest so that relative to the car

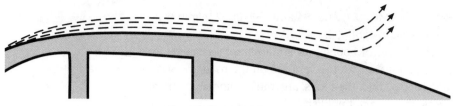

Fig. 62. Boundary layer.

it has the same velocity as the car. This gradual lagging gives rise to a gradient, the boundary layer. How thick is this layer? It is usually very thin—like a sheet—and indeed, as the speed of the car increases, it gets even thinner. It gets thicker, however, as it approaches the rear of the car. The frictional force due to the layers of air close to the surface is referred to as *surface friction drag*. It acts in a direction tangential to the surface.

When the friction between the layers is not high, the layers slide over one another quite easily. In this case we have what is called *laminar flow* (smooth flow), which occurs only at relatively low speeds over a car. The transition from laminar flow to turbulent flow is an important part of aerodynamics and considerable study has gone into it. Osborne Reynolds was one of the first at the turn of the twentieth century to study it in detail, and as a result we now have what is called *Reynolds numbers*. These numbers depend on the viscosity, the density, and the velocity of the fluid. If the Reynolds number is between 0 and 2000 the flow is laminar; if it is over 3000 the flow is turbulent. The region in between is a transition region where the flow can change back and forth. (See fig. 63.)

Form Drag

As we saw earlier, form drag depends mainly on the shape of the car. Frictional forces produce pressure dif-

Fig. 63. Flow lines over a car.

ferences at right angles to the surface, and if we add up all these pressures over the area of the car we get the total form drag force on the car.

Form drag also depends on the splitting of the streamlines and the wake behind the car. It is important to reduce splitting as much as possible and also to keep the wake to a minimum. The energy that goes into creating the wake is taken out of the forward motion of the car and therefore reduces the car's horsepower.

Bernoulli's Theorem

One of the most important relations in aerodynamics is referred to as Bernoulli's theorem. It was formulated by Daniel Bernoulli in the eighteenth century. Bernoulli, who came from a long line of Swiss mathematicians, is best known for his book on the flow of fluids. He was particularly interested in water and other fluids, but his ideas also apply to air.

In 1738 he showed that as the velocity of a fluid increases, its pressure decreases. Mathematically, we can state this as

$$p + \rho \, v^2/2 = \text{constant},$$

where p = air pressure, ρ = density of air, v = velocity, and $\rho v^2/2$ is the dynamic pressure. We can see from this that *as dynamic pressure, which depends on velocity, increases, air pressure must decrease, and vice versa.* This result is strictly applicable only to cases where the viscosity is zero, but outside the boundary layer we can assume this is the case.

This increase in pressure is the reason an airplane wing produces lift. It is designed so that the velocity of air across the top is greater than across the bottom. If the velocity is greater, the pressure is less (fig. 64). This means the pressure beneath the wing is greater, and this in turn produces the lift which allows the plane to fly.

We have the same situation in a curving baseball. A spin is put on the baseball so that the velocity of air varies from point to point around it. This creates pressure differences that tend to change the direction of the ball.

Drag Force and Drag Coefficient

Both the drag force on a car and the coefficient of drag, c_d, can easily be calculated. The drag force F_d is given by

Fig. 64. Flow lines around a wing.

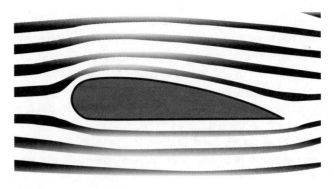

$$F_d = \rho v^2 A_f c_d / 2,$$

where A_f is the frontal area of the car. It is convenient to substitute an average value for the pressure of the air and change the units appropriately. When we do so, we get

$$F_d = (v^2 A_f c_d)/400.$$

Here the velocity v is in mi/hr, the frontal area A_f in square feet, and c_d is dimensionless.

Let's consider a few cases. We know that c_d's range from .5 to .3, so we'll consider a range of velocities and take the frontal area to be 18 square feet.

At 40 mph the drag force in pounds is

$$F_d = (40^2 \times 18 \times .5)/400 = 36 \text{ lb.}$$

Similarly, for several other speeds and c_d's, see table 7.

Note that if we measure the drag force in a wind tunnel, we can use this formula to calculate c_d:

$$c_d = 400 \, F_d / v^2 A_f.$$

As an example, assume F_d is 50 lb at a speed of 50 mph. We get

$$c_d = (400 \times 50)/(50^2 \times 18) = 0.44.$$

It is convenient to transform this formula so that it gives the drag in equivalent horsepower. We will then know how much horsepower is used up in overcoming drag, and therefore how serious it is in relation to fuel economy. We have

$$F_d(HP) = (v^3 A_f c_d)/150,000.$$

Table 7. Drag force at several speeds for various c_d's

c_d	v (mph)	F_d (pounds)
.3	40	21.6
	50	33.8
	60	48.6
	70	66.2
	80	86.4
	90	109.4
.4	40	28.8
	50	45.0
	60	64.8
	70	88.2
	80	115.2
	90	145.8
.5	40	36.0
	50	56.3
	60	81.0
	70	110.3
	80	144.0
	90	182.3

At 70 mph with a c_d of .4, this gives F_d = 16.5 hp. If your vehicle has a horsepower of 200 this may not seem very significant, but at 100 mph we get 48 hp, which obviously is significant. (See table 8.)

As it turns out, aerodynamic drag is not the only force acting against the forward motion of the car. Rolling resistance is also important. In most cases it is much less than form drag but it does have an effect. It is relatively difficult to calculate, so please refer to figure 65 (p. 150).

Frontal Area of Car

In the formula for aerodynamic force and coefficient of drag, one of the important components is the frontal

Table 8. Horsepower equivalent of form drag for several speeds for various c_d's

c_d	v (mph)	F_d (HP)
.3	40	2.3
	50	4.5
	60	7.8
	70	12.3
	80	18.4
	90	26.2
	100	36.0
.4	40	3.1
	50	6.0
	60	10.4
	70	16.5
	80	24.6
	90	35.0
	100	48.0
.5	40	3.8
	50	7.5
	60	13.0
	70	20.6
	80	30.7
	90	43.7
	100	60.0

area of the vehicle. This area should therefore be kept as small as possible if the drag force and c_d are to be low. The frontal area can be determined using a laser and a photograph of the front of the car. To a first approximation, we can use 80% of the height times the width.

Since frontal area and c_d are both of particular importance, it is worthwhile to consider their product, called the *figure of merit*. The figure of merit gives a better comparison between cars. Since frontal area and c_d

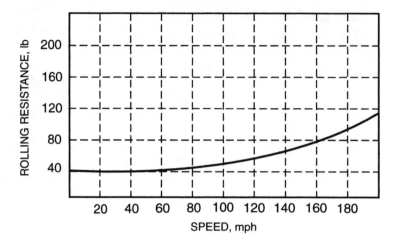

Fig. 65. Rolling resistance versus speed.

can be traded one for one, it is possible that a vehicle with a low c_d can have a relatively large frontal area, and vice versa. So, one of the numbers does not tell the entire story. Their product, however, does.

Frontal areas have decreased considerably in recent years. Cars of the 1950s typically had frontal areas of 25 to 26 square feet. Today, most vehicles have frontal areas of 18 square feet or less. With a frontal area of 18 square feet and a low c_d of .3 we have a figure of merit of 5.4.

Reducing Drag

The purpose of applying aerodynamics to cars is, of course, to reduce drag. How low can it get? As we saw earlier, an airplane wing has a c_d of .05, but this is of little help to us in designing passenger cars. It has long been known, however, that the teardrop, or fish, shape as shown in figure 66 is the ideal. It has a drag coefficient of about .03 to .04.

The ideal situation with airflow around a car is that

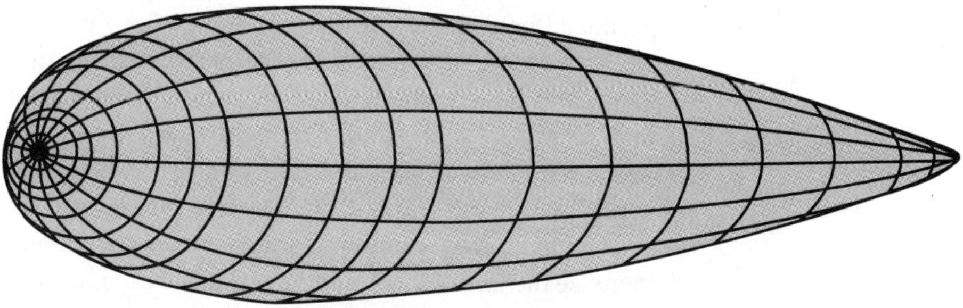

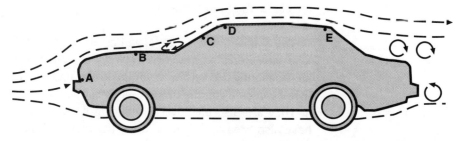

Fig. 66. The ideal aerodynamic body—a teardrop shape.

the streamlines follow the car's contour from the front to the back without breaking away. In practice, this rarely happens. Even with the teardrop, there is turbulence near its end. Let's consider the flow in more detail, using the diagram in figure 67. The onrushing air splits at point A, with some flowing over the car and some under. The airflow over the car usually splits slightly at B (just before the windshield) causing turbulence. It rejoins at C, then rushes along the top of the car (D), where it may split at E and create more turbulence, or it may leave the car smoothly. In the case of a fastback, the flow can remain attached to the end of the body, leading to a small wake.

The wake behind the car is of considerable importance. If the flow separates too abruptly, a vacuum is

Fig. 67. Flow lines over and under a car. Note areas of turbulence in front of windshield and behind car.

produced behind the car that applies a force on it. As we saw earlier, a way around the problem of a large wake, and also the sharp tail of the teardrop, was found by Wunibald Kamm and others of Germany in the 1920s. He showed that the rear end of the teardrop could be cut off without significantly increasing the aerodynamic drag. Cutting off this region does, indeed, increase the form drag slightly, but there is a loss of skin friction because of the reduced body area. It is critical, however, that the cut be made directly behind the rear wheel axle.

The area of the cut is referred to as the *base area,* and it should be kept as small as possible. The technique of cutting off the end section is sometimes referred to as *bobtailing.*

Keeping the overall shape as close as possible to the teardrop shape is one way of reducing drag, but there are others. For years, little attention was paid to the underside of cars, but we now know that considerable drag is introduced here. The airflow in this region is very complex, and there is considerable turbulence. The car moves very close to the ground, restricting airflow and increasing pressures. We will look into the consequence of these effects later. In general, the airflow in this region is affected by the distance between the ground and the underside, the width and length of the car, and, of most importance, the roughness of the underside.

The airflow under the car is particularly important in race cars; therefore, low curved "bumpers" called dams are frequently used. They divert the air around the car, thereby decreasing the amount of air passing beneath it. They are also effective in reducing lift. Belly pans have also occasionally been used, but they have been forbidden in most racing.

Interference and Other Forms of Drag

Many things besides the shape of the car and the underside of it can cause drag. Small protuberances such as radio aerials may seem insignificant, but they do create some drag. Outside rear-view mirrors, windshield wipers, door handles, and so on also produce drag. Indeed, because of interactions with the airflow of the car itself, their overall contribution is usually greater than the sum of their individual contributions. They give rise to *interference drag.*

Another source of considerable drag is the wheels. The pressure around the surface of the wheel differs from point to point, producing drag. The wheel spins the air with it and produces turbulence in the form of vortices. These vortices interact with the vortices created by the overall motion of the car. The drag coefficient of a wheel exposed to airflow is .45, so inserting the wheel in the car body is advantageous, but there are complications. (See fig. 68.)

Finally, there is internal drag caused by the air that flows through the car. All engines require forced air to

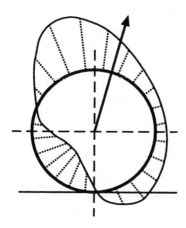

Fig. 68. Pressure lines around a tire.

keep them cool, and this air must be brought in from the front. It is important, in fact, that there exists a pressure difference between the air at the front of the car and that near the rear. Overall there is little that can be done to reduce this drag, and it is a necessary evil.

Aerodynamic Lift and Downforce

Lift is normally of little importance in passenger cars as their speed is usually too low to produce much lift. It was noticed early on that something strange happened at high speeds: the car seemed to be lifting off the ground. We now know that lift can be serious, particularly in racing cars. It has a serious effect on the control and handling of the car.

Lift occurs because the airflow over the top of a car is faster than across the bottom (fig. 69). This occurs to some degree in all cars. As the speed increases, the pressure decreases, according to Bernoulli's theorem. The top of the car therefore has a lower pressure than the bottom, and the result is a lifting force. Comparing a car to an airplane wing, we see the similarity.

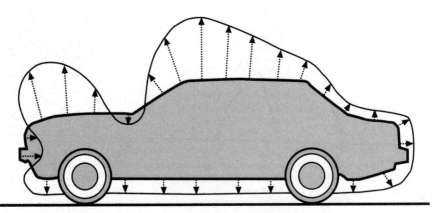

Fig. 69. Pressure lines over and under a car that give rise to lift.

The lift force is given by

$$F_l = \rho v^2 A c_l / 2,$$

where A is the area of the underside of the car and c_l is the coefficient of lift. In appropriate units this becomes

$$F_l = v^2 A c_l / 391,$$

where v is in miles per hour and A is in square feet, and c_l is a number, like c_d. Lift coefficients can vary from 0 up to 2 or more. They depend on the shape of the nose of the car, the overall styling, and the angle of the nose. Let's assume c_l is .5 and A is 36 sq ft. In this case we get the following:

v (mph)	F_l (pounds)
50	115.0
60	165.6
70	225.4
80	294.4
90	372.6
100	472.0
120	662.4
150	1035.0

We see that the lifting force can be substantial if the speed is very great and in the case of racing cars it can be several hundred pounds.

One of the major effects of lift is to reduce traction, which can have a serious effect on the speed of the car, and on the maximum cornering force. Not only would we like to reduce lift, but we would also like to increase the downforce, in other words, the force pushing the tires onto the road. But we want to do this without increasing the weight of the vehicle. There are two ways of doing this: with spoilers and with negative lift devices such as wings.

Spoilers were first used in Ferrari racing cars in the early 1960s and are sometimes referred to as *Ferrari spoilers.* Spoilers interfere with the airflow pattern over the upper back surface of the car. They produce a high pressure at their upper edge that is projected backward to the area behind the rear window. With an increase in pressure here, the lift is decreased. Strangely, while spoilers were originally used for reducing lift, it was later found that they also reduce drag. (See fig. 70.)

Wings are usually mounted near the rear end of a racing car. They are, basically, an airplane wing mounted upside down. It is well known that an airplane wing produces upward lift—enough to lift the airplane off the ground. Mounting it upside down changes the direction of the force, making it very effective in producing a downforce. We can use the same formula we used for calculating the lift on a car for determining the lift of a wing. The area in this case is the area of the wing in square feet.

To see what kind of a downforce we can expect, let's substitute some numbers. We'll assume c_1 is .85 and take the area to be 30 square feet. For a velocity of 120 mph we get $F_1 = 939$ lb. But this isn't the end of the story. Negative lift produces drag, not only form drag but a new type referred to as *induced drag.* Form drag is cal-

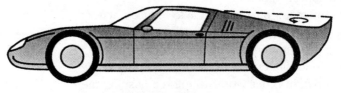

Fig. 70. A spoiler.

culated from the usual formula at the beginning of the chapter. Induced drag is determined from

$$F_i = (c_i^2/a)v^2/4103,$$

where a is the aspect ratio = $span^2/A$.

Stability of the Car

In the last section we dealt with lift, which can be considered as up-down stability. But directional stability is also important. To understand it, we must consider slip angle and yaw, quantities that are not directly related to aerodynamics. But aerodynamics certainly affects stability at high speeds.

We considered slip angle earlier. It is the angle between the direction that the tires are rolling and the direction in which they are pointed. The flexibility of rubber allows this misalignment. The side-slipping tire produces a lateral force perpendicular to it, and the magnitude of this force depends on the size of the slip angle and on the normal force pushing the tire onto the road.

When the driver turns the steering wheel, a slip angle develops, and it in turn produces a lateral force that pushes the car in a particular direction. This force acts on the front tires, which are some distance ahead of the center of gravity of the car. The result is a torque. The car therefore begins to rotate around its center of gravity. This rotation is called *yaw.* When this happens, the rear tires also develop a slip angle and a lateral force. But this force is behind the center of gravity and so acts to counter the front force.

Looking closer, however, we see that the slip angle in the front tires is greater than the backs, therefore the car continues to yaw. In fact, if the driver does not take evasive action, this yaw will cause the motion of the car to become unstable, and it will spin out. The driver must readjust the steering wheel to avoid this.

In some cases it may be necessary to modify something in the car or tires to assure that the torque generated by the front tires differs from that generated by the rear tires. Different inflation or different-sized tires are commonly used.

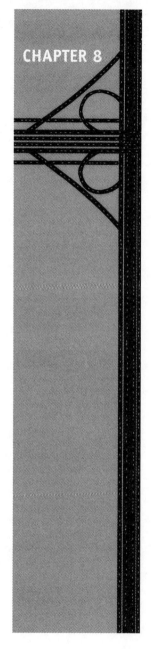

CHAPTER 8

A Crash Course

The Physics of Collisions

Two cars speed down the highway in opposite directions. One of the drivers begins to nod off. He crosses the yellow line, heading for the other driver. The second driver jams on his brakes and swerves but he can't avoid the crash. The two cars hit head-on. The first driver is hurled through his car's windshield and ends up on the hood of the other car, dead. The second driver hits the steering wheel and is injured, but unlike the first driver he survives the crash. The two cars are the same weight and were traveling at roughly the same speed. It might seem strange that one driver would be killed while the other survives. After all, at first glance conditions appear to be the same for both drivers. When we look closer, though, we see that there is a *big* difference. One was securely attached to his seat by his seat belt, the other was not.

I'll admit, maybe this story is a little gruesome, but I think it makes the point. Seat belts do save lives. We know that, according to the law of inertia, an object in motion will continue in motion, with the same speed and direction, unless acted upon by an outside force.

The driver who died in the crash was not wearing a seat belt and was therefore not connected to his car. When the collision occurred he continued with the speed the car had before the collision, first hitting the windshield and then continuing to fly through it to the hood of the other car. The force of the impact with the windshield and his encounter with the hood of the second car slowed him down and stopped him, but by then it was too late to save his life. The driver who survived was securely attached to his car by a seat belt, so he experienced the same motion and deceleration as the car and this saved his life.

Crashing into a Fixed Solid Object

Physics is important in understanding car crashes. As we will see, many different situations can occur. One of the simplest is a car crashing into something solid and immovable such as a tree or brick wall. We can easily determine the force with which the car hits the wall, and this is also the force the driver's body will experience. Let's assume the car is going 50 mph and is brought to rest in 0.04 second. The two physical concepts we need to solve the problem, momentum and impulse, were introduced earlier. Momentum is mass times velocity, or mv, and impulse I is

$$I = Ft,$$

where F is force and t is the time over which the force acts. We can relate them using Newton's second law,

$$F = ma = m(v - v_0)/(t - t_0),$$

where v_0 is the original velocity and t_0 is the original time, which we can take to be zero. Rewriting this, we get

$$F = \Delta mv/t$$

or

$$Ft = \Delta mv.$$

If the weight of the car is 3000 pounds, its mass is 3000/32, or 94 slugs (where a slug is the unit of mass), and its velocity (50 mph) is 73.3 ft/sec. Substituting these into our formula gives

$$F = 94(73.3)/.04 = 172,255 \text{ pounds.}$$

Needless to say, this is an incredible force, and it could do a lot of damage. Imagine trying to support this kind of weight. To get a better feel for it, let's convert it to g's. We can do this using $a = F/m$. Substituting, we get 1833.3 ft/sec^2, which corresponds to a force of about 57 g's. Could a human body take this many g's and survive? Published safety standards say that it is possible for a person to survive decelerations up to 80 g's. What is critical, though, is the period of time over which it takes place: it must be exceedingly short. Furthermore, we assume that the glass and sharp metal pieces created in the crash don't kill the passenger, and there's a good chance that this can happen. Later we will see that we can define a severity index (SI) that takes both the acceleration and time into account. It gives a good estimate of the probability of survival.

The Head-on Collision

One of the most common and the most lethal of all collisions is the head-on collision. It is a one-dimensional problem and therefore relatively easy to solve. A variation of it is the case where both cars are going in the same direction and one car is rear-ended (fig. 71). The mathematics for this case is no different; only the sign changes.

First, we need to compare the momentum before and after the collision. It is given by the principle of conservation of momentum, which says that the total momentum before a collision is equal to the total momentum after. Mathematically it can be expressed as

$$m_1v_1 + m_2v_2 = m_1V_1 + m_2V_2,$$

where m_1 and m_2 are the masses of the two cars, v_1 and v_2 are their velocities before the collision, and V_1 and V_2 are their velocities after the collision.

If we know the velocities before the collision we are still left with two unknowns, so we can't solve the problem. But there are two cases that we can solve. They are:

1. The perfectly elastic collision between the two vehicles, where they rebound from each other.
2. The perfectly inelastic collision, where the two vehicles stick together after the collision.

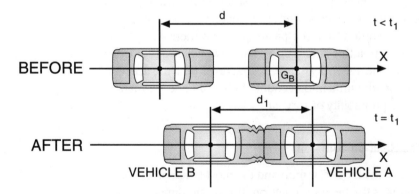

Fig. 71. Two cars in a head-on collision.

The first of these is of little interest because it never occurs in practice, but it's easy to solve so we'll look at it briefly. The second occurs occasionally and is of interest.

Let's begin with the perfectly elastic collision. We have the momentum equation above, and we also have another equation. In a collision of this type, kinetic energy is conserved, so we have

$$\tfrac{1}{2}m_1 v_1^2 + \tfrac{1}{2}m_2 v_2^2 = \tfrac{1}{2}m_1 V_1^2 + \tfrac{1}{2}m_2 V_2^2.$$

We can write these two equations as

$$m_1\,(v_2^2 - v_1^2) = m_2\,(V_2^2 - v_2^2)$$

$$m_1\,(v_1 - V_1) = m_2\,(V_2 - v_1).$$

Dividing the first by the second gives

$$v_1 - v_2 = V_2 - V_1.$$

This tells us that the velocity of approach of the two vehicles before the collision is equal to the velocity of separation after the collision. This, of course, is what is expected in a perfectly elastic collision. We can also determine V_1 and V_2, the velocities after the collision, but they are usually of little interest so we'll ignore them.

Turning now to the more interesting case of the perfectly inelastic collision, we have the two cars sticking together after the collision. Our momentum equation is

$$m_1 v_1 + m_2 v_2 = (m_1 + m_2)\,V,$$

where V is the common velocity after the collision. We

do not have an energy equation this time, as energy is not conserved. As an example, let's assume that two vehicles, one of 3000 pounds and another of 4000 pounds, collide head-on and the 3000-pound car has a velocity of 50 mph and the other one a velocity of 60 mph in the opposite direction. The respective masses are 125 and 93.75 slugs, and the speeds in feet per second are 73.5 and 88.2. Substituting, we find V, the velocity of the two cars after the collision, is 42 ft/sec in the direction that the 4000-pound car was traveling initially. They won't have this velocity for long, however, because of friction.

In reality, most collisions are somewhere between the above two cases. In other words, they are neither perfectly elastic nor perfectly inelastic. To deal with them we need the *coefficient of restitution*. Designated by e, it is defined as

$$e = (V_1 - V_2) / (v_2 - v_1) = \text{velocity of separation/} \text{velocity of approach.}$$

In the case of a collision of two cars it is difficult to determine e exactly. It ranges between 0 and 1, with the two extremes being the two cases we dealt with above. The coefficient of restitution of a bouncing ball is relatively easy to determine. If we let it fall from a height h_1, and it bounces to a height h_2, the coefficient of restitution is h_1/h_2. We obviously can't measure the coefficient of restitution of a car collision in the same way, so we usually have to make estimates.

In terms of the coefficient of restitution, the velocities of the two vehicles after the collision are relatively easy to calculate:

$$V_1 = [(m_1 - em_2) v_1 + m_2 (1 + e) v_2] / (m_1 + m_2)$$

$$V_2 = [m_1 (1 + e) v_1 + (m_2 - em_1) v_2] / (m'_1 + m_2).$$

In the above discussion we saw that kinetic energy is conserved in a perfectly elastic collision. In all other cases this is not true. Indeed, if we calculate the total kinetic energy before the collision and compare it to the total kinetic energy after the collision, we will see that they are not equal. In fact, a lot of kinetic energy will appear to be lost. Where did it go? At first glance it appears that energy is not conserved, but we know that can't be true; it has to be conserved. What, then, has happened to it? It was not lost, but was transformed into different types of energy, with much of it becoming heat energy. Both of the cars were no doubt crumpled and smashed in the collision, and it takes work or equivalently, kinetic energy, to do this. Furthermore, there was likely some sound energy and perhaps a little radiant energy during the collision, and it all adds up.

Collisions in Two Dimensions

Many collisions do not occur on a highway and are not head-on or rear-end collisions. Collisions at intersections, in fact, are more common than highway collisions. The physics of this case is more complicated because we are now dealing with two dimensions. But again the problem can be solved. The only difference is that we will need trigonometry. Taking the horizontal axis as the x-axis and the vertical one as the y-axis, we project all velocities onto these two axes (fig. 72). This is done rather easily. In figure 72, a velocity v is $v \cos \theta$ along the x-axis and $v \sin \theta$ along the y-axis. After projecting all velocities onto both axes, we assume that momentum is conserved in the x and y directions and the problem can be solved as above.

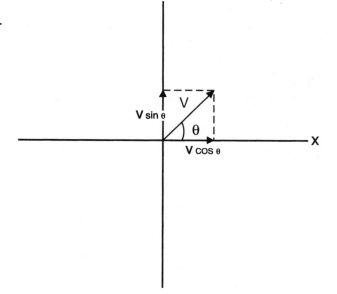

Things can get quite complicated in real life. A car is not a point, and the collision at an intersection can occur in many ways. For example, the front of one car could smash into the front of the other car, or it could collide with the center of the side, or near the rear end. In the case of a collision near the front or rear, the first car would impart a twist, or torque, to the second car, which would have to be dealt with in an exact calculation. In addition, one of the other drivers may see the collision coming at the last moment and swerve to avoid it. In this case they would not collide at right angles. You could still deal with this case using the above approach, but the angles would be different. Something else that is significant is whether the vehicles have their wheels locked after the collision. This can occur if deformations in the collision are sufficient to prevent the wheels from rotating. This would have to be taken into consideration in a detailed calculation. (See fig. 73.)

Nowadays computers are used extensively in simulating collisions, and this helps with the mathematics. In advanced simulations where factors such as the geometry of the front of the vehicles and the exact position of the collision are taken into consideration, the mathematics can become quite complicated and computers are usually needed. Several computer programs have been designed specifically for dealing with car

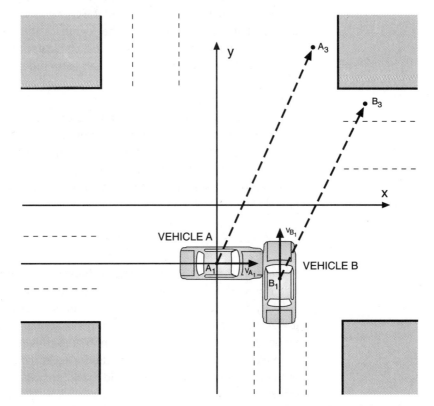

Fig. 73. Collision at an intersection. Arrows show the directions of the cars after the collision.

collisions. Information about them can be found on the Internet at www.e-z.net.

Reconstructing Accidents

When a collision occurs at an intersection, all we have to go on are the final positions of the cars and the skid marks. It is important in many cases, particularly those that go to court, to be able to reconstruct the accident. This means determining the initial velocities of the two cars before the collision. Of particular interest in many cases is whether one or the other of the cars was speeding. A vital clue, of course, is the length of the skid marks. But there is a problem. A car must have its brakes fully locked to leave skid marks. We can therefore determine only its speed after the driver has applied the brakes and the wheels have locked up. The actual initial speed is therefore going to be slightly greater than the speed we calculate. If you need a better value you will have to make an estimate of the reaction time of the driver and the time for the brakes to lock up.

If you have the length of the skid marks, the original speed can be obtained from

$$v = 5.5(\mu l)^{\frac{1}{2}},$$

where μ is the coefficient of friction between the tires and the road, l is the length of the skid marks, and v is velocity in miles per hour. As an example, assume $\mu = .7$ and the length of the skid marks is 40 feet. The velocity is then $5.5 (.7 \times 40)^{\frac{1}{2}} = 30$ mph. It is convenient to make up a table using this formula for several different coefficients of friction and skid mark lengths. (See table 9.)

In table 9 we assume the car comes to rest as a result

Table 9. Skid mark lengths for various initial velocities

Skid Mark Length (feet)	μ	Initial Velocity (mph)
40	.5	24.6
	.6	27.0
	.7	29.1
	.8	31.1
50	.5	27.5
	.6	30.1
	.7	32.5
	.8	34.7
60	.5	30.1
	.6	33.0
	.7	35.6
	.8	38.1
80	.5	34.8
	.6	38.1
	.7	41.2
	.8	44.0
100	.5	38.9
	.6	42.6
	.7	46.0
	.8	49.0
120	.5	42.6
	.6	46.7
	.7	50.4
	.8	53.9
140	.5	46.0
	.6	50.4
	.7	54.4
	.8	58.2
160	.5	49.2
	.6	53.9
	.7	58.2
	.8	62.2

of friction between the road and the tires. In many cases, however, the skid marks end with a collision, and we can't determine the initial speed. If the initial speed is known, however, the length of the skid marks will give you the speed with which the first car hit the second. Although it is not nearly as accurate, the initial speed of a car can also be obtained from an estimate of the amount of damage it causes to the second car.

Also important in reconstructing an accident is the time over which the impact takes place. It is usually very short—only a fraction of a second. As an example, assume that a car collides with another car at rest at an intersection, and we have determined that the second car was pushed two feet by the impact and is crumpled by one foot. The first thing we need is the average velocity during the impact. If we know the velocity of impact we can estimate it, since we know the final velocity is zero. If the velocity at impact was, say, 10 mph, then we can take the average velocity to be 5 mph (7.35 ft/sec). The time of impact is then

$$t = 3 \text{ ft}/7.35 \text{ ft}/\text{sec} = 0.4 \text{ sec.}$$

Severity Index

What we would really like is a measure of how serious an accident is—in other words, a number that can tell us the probability of a fatality or serious injury. And we do, indeed, have such a number, known as the *severity index*. We will designate it as SI. It is defined as

$$\text{SI} = (a/g)^{5/2} t,$$

where a/g is the g force on your body at impact, and t is the time of impact.

To illustrate its use, let's consider another case. Suppose a car starts across an intersection but is suddenly stopped by a child running out in front of it. In the meantime, a half ton truck coming into the intersection from a perpendicular direction fails to see the light and slams into the car without braking. After the accident, police determine that the skid marks are 83 feet long and the coefficient of friction between the road and tires is .7. We'll assume the collision is inelastic and the two vehicles move off together after the collision.

The weight of the car is 3050 pounds and the driver weighs 150 pounds for a total of 3200 pounds. The truck weighs 3750 pounds and the driver 200 pounds for a total of 3950 pounds. Let's begin with the retarding force of the car. It is given by

$$F = \mu mg = .7\ (3200) = 2240 \text{ pounds.}$$

From the conservation of momentum for an inelastic collision we have

$$m_1 v_1 + m_2 v_2 = (m_1 + m_2)\ V.$$

With $v_1 = 0$ we get v_2 (the velocity of the truck) = 7150/3950 V. The deceleration during the collision is

$$a = -F/m = -10 \text{ ft/sec}^2.$$

Now using $V^2 = 2as$, where s is the length of the skid marks, we get the velocity of the truck and car immediately after the collision as 40.7 ft/sec or approximately 28 mph. The velocity of the truck just before it struck the car is therefore 1.81(28) = 50.6 mph.

To proceed further we have to make some rough approximations. We need the crumple distance for the car and the time over which this crumpling took place. Both of these are difficult to determine exactly, and I won't go into the details. We'll use the values we got earlier in our example where a car struck a wall. The crumple time was 0.04 second and the g force was 57. In that case,

$$SI = (57)^{5/2} \, (.04) = 981.$$

It is generally assumed that if the SI is less than 1000 the person in the crash will survive. In this case it is obviously close.

Crashworthiness

Crashworthiness is a measure of how well a car survives a crash. In particular, it gives a measure of the degree of injuries we could expect. What we want, of course, is a high degree of crashworthiness; in other words, we want the injuries to be minimal. In analyzing crashworthiness we therefore have to look into whether something was lacking, something that would have made the collision less injurious. Specifically, given what we know about the injury mechanisms of accidents and having some means of estimating the severity of the crash, we can determine whether the occupants of the car could have come out of the collision with fewer injuries.

It's important to note that crashworthiness is not the same as vehicle safety. In dealing with crashworthiness we are assuming the collision has already occurred and we are not concerned with whose fault the accident was or whether it could have been avoided. Vehicle safety depends on many factors that may be important in avoiding an accident, things such as ABS, good steering and handling characteristics, type of tires, and so on. But a relatively safe vehicle could have poor crashworthiness. Even if it were equipped with all the appropriate features for avoiding an accident it could still have features that could cause more injuries than necessary.

Things that are important in relation to crashworthiness are seat belts, air bags, side-impact protection, crumple zones, head rests, and interior padding. A crumple zone, in case you're not familiar with it, is a zone in front of the car that is designed to fold up or crumple to absorb some of the collision's force. A cage, or solid metal frame, encases the driver behind this region. The crumple zone takes the pressure off other areas of the car.

Most of the above features are now available in cars, but some cars have them to a larger degree than others. In considering crashworthiness, all types of impacts must be considered, namely head-on collisions, side collisions from either side, rear-end collisions, and various types of oblique and glancing collisions. Features such as the collision of passengers with the dashboard and steering wheel must also be considered. Collapsible steering wheels, for example, make a car more crashworthy. Also, things such as ejection from the car and danger of fire must be considered.

How do we determine crashworthiness? This question takes us to crash tests.

Crash Tests

A car suddenly swerves in front of you. You jam on your brakes and brace yourself. A collision is unavoidable. How bad will it be? To a large degree that will depend on the safety rating of your car. In the United States there are two major safety ratings, one by the government and the other by major insurance companies. The first is conducted by the National Highway Transportation Safety Administration (NHTSA), the second by the Insurance Institute for Highway Safety / Highway Loss

Data Institute (IIHS/HLDI). The government rating is based on stars, with five stars being the best. Two crash dummies, fitted with sophisticated data-gathering devices, are belted in the front seat, and the car hits a nondeformable barrier at 30 miles per hour. The tests are referred to as NCAP (New Car Assessment Program).

The devices within the dummies measure the acceleration of the head and chest and the pressure on the thighs as the car strikes a barrier. The numbers that come out of the devices are fed into a computer program. Several stars are then assigned to the car, depending on the number that comes out of the computer. Five stars is maximum; in this case the car has exceeded the federal crash standards, and at the speed of the collision there is only a 10% chance of serious injury. Four stars means the vehicle has also exceeded the federal crash standards, but there is now a 10% to 20% chance of a serious injury. A three-star rating means the car has passed the standards but is marginal. Anything below this fails the test. As in the government tests, dummies with sensing devices are also used in the IIHS/HLDI tests. In this case the cars are rated as good, acceptable, marginal, and poor. The first three exceed the federal crash standards, the last does not.

In the government tests the devices are designed mainly to check on the safety of the seat belts and airbags. They also assess the crash characteristics of the body structure. The Europeans check their vehicles with tests similar to the IIHS/HLDI test. In recent years, side-impact tests have also become increasingly common, and more recently rollover tests have been conducted. Both of these are important in a collision, but they are still in the early stages of testing and I will say little about them.

Both the NHTSA and IIHS/HLDI tests have done a lot to increase the safety of modern cars. Almost all cars are now equipped with airbags, crumple zones, and so on, making them much safer than the cars of the 1950s, 1960s, and later. A check over the past few years also shows that NCAP scores have improved significantly for most vehicles, so it is obvious that manufacturers are taking them seriously.

The crash ratings for cars can be found in many publications and on the Internet at www.highwaysafety.org. Table 10 shows a few of them. It is easy to see from the table that the two tests do not always agree.

Although the tests are, without a doubt, helpful, they are flawed. You can only test so much by crashing a car into a nondeformable barrier at 30 miles per hour, and most accidents that occur are not of this type. Indeed, most crashes are between cars, with each car having a crumple zone. Because of this, the European tests are conducted using a deformable honeycomb barrier, which is closer to the conditions in a real crash.

Table 10. Crash ratings for various 2002 vehicles

Model	NHTSA (Driver/passenger)	IIHS/HLDI
Audi A6	★★★★/★★★★	Acceptable
Buick LeSabre	★★★★★/★★★★★	Good
Chevy Impala	★★★★★/★★★★★	Good
Honda Civic Sedan	★★★★★/★★★★★	Acceptable
Infiniti QX4	★★★★/★★★★★	Marginal
Isuzu Rodeo	★★★★/★★★★	Poor
Lincoln LS	★★★★★/★★★★★	Good
Mercury Sable	★★★★★/★★★★★	Good
Ford Taurus	★★★★★/★★★★★	Good
Oldsmobile Aurora	★★★★/★★★★	Good
Plymouth Neon	★★★★/★★★★	Marginal
Saturn LS	★★★★/★★★★★	Acceptable

The major problem with the tests is that real accidents differ significantly. Many types of collisions occur, and all of them can't be tested. The tests do, however, give a good indication of the crashworthiness of the vehicle. Airbags, seat belts, crumple zones, and so on have definitely increased our chances of survival in a collision, but also of importance is the ability to avoid an accident. This depends on several factors, including what the handling characteristics of the car are, how sharp the car can turn, how fast it can stop, and how well it sticks to the road. ABS, traction control, the suspension system of the car, comfort, and features that allow fatigue-free driving on long trips all contribute to the safety of the car. But perhaps most essential is the skill and ability of the driver. For accident prevention it is important that the driver have a good foundation of skills.

So you've looked up the crash rating of your car and you're pleased. It has a five-star rating. Before you get too smug you should consider the following. These ratings give good information about how much damage you could expect for a car hitting an immovable barrier. But the five stars won't help much if you hit a large truck head-on. By the laws of physics, your car is going to come out second best. The momentum of your car is likely to be considerably less than that of the truck.

In the past few years there has been considerable worry about the large number of SUVs and pickups on the road as compared to passenger cars. Studies have shown that the people riding in a car (even if it has a good crash rating) are four times more likely to be killed if they are struck by an SUV or pickup. Furthermore, they are eight times more likely to be killed if hit on the side by an SUV or pickup. The reason is, of course, the difference in weight. The Lincoln Navigator, for example, weighs 5500 pounds, compared to about 3000 for most cars. But weight is not the only problem. SUVs and trucks ride considerably higher than cars, and their bumpers are frequently higher than those of a car. Furthermore, the frames of SUVs and pickups are usually stiffer and stronger than those of a car.

Collision Protection

Most cars are now equipped with many protection devices. One of the oldest is seat belts, and indeed, they are critical in any collision. The seat belt makes the person part of the car, so that he or she is decelerated at the same rate as the car and not thrown through the windshield. Shoulder straps are also useful in that they help the upper part of your body slow down at a slower rate, but they can place a tremendous pressure on your chest.

Seat belts are, however, not enough and should be used in conjunction with airbags, and since 1999 airbags have been mandatory in all new cars. When seat belts are used in conjunction with airbags, injuries are usually reduced considerably. When a car is stopped suddenly, an impulse $I = Ft$ is generated according to the momentum lost. If the time of the impact is exceedingly short, as it usually is, the force can be very high, as we saw earlier. Airbags cushion you from this force. But under certain circumstances they can be dangerous. A child's safety seat, for example, should never be installed directly behind an airbag. Also, small people, who frequently pull the seat as far forward as possible, are in some danger. Furthermore, if you aren't wearing a seat belt, airbags can injure you more than the collision. Despite these problems, airbags are a valuable addition to a car, and they have saved thousands of lives.

A crumple zone in the front of your car is also critical. If your car is equipped with a good crumple zone it can considerably increase the time of impact and thereby decrease the magnitude of the force. Other safety features of importance are collapsible steering wheels, side airbags, and a breakaway feature within the car. Seats are also critical. They cushion the passengers in a collision and therefore must be soft. They are particularly important in rear-end collisions. Something else of significance in a rear-end collision is a head rest, and many cars are now equipped with them. They can save you from considerable whiplash or even a broken neck.

What are the most common types of injuries that occur in an accident? One of the worst is head injuries, with concussion being common. In a collision the brain slams against the skull and the resulting squeezing effect can cause a concussion. Harsh chemical reactions are unleashed in the process, and it can take considerable time to overcome the effects of a concussion. Neck injuries are also common as whiplash usually occurs when a car is suddenly stopped.

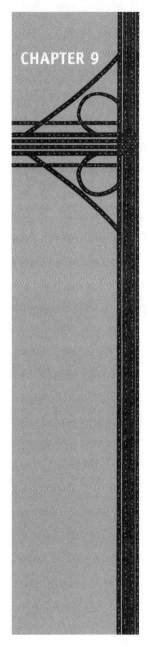

CHAPTER 9

Checkered Flags

The Physics of Auto Racing

The crowd was on its feet. Dave Pearson and Richard Petty were battling it out in the 1976 Daytona 500. First, Petty was in the lead, then Pearson. Then, on the last lap, while coming into the last turn to the checkered flag, they collided. Both cars flew off into the infield. Fans in the stands watched in awe as each man tried to get his car moving. Both were only a short distance from the finish line. Neither car would start, but Pearson persevered. Pushing on the starter of his disabled car, he was able to nudge his car over the finish line.

The Daytona 500, as all race fans know, is part of NASCAR. Both modified and unmodified stock-car races are held. Today the cars resemble stock cars in outer appearance, but beneath their exterior they are quite different from ordinary stock cars. Much of the difference has been incorporated for safety reasons, however. Indeed, although stock cars are the most common type used in NASCAR, many other types are now used. Even trucks are now raced.

In the other big race—the Indianapolis 500—a completely different type of car is used. Smaller and more streamlined, it is referred to as an Indy car. The European equivalent is the formula 1 and formula 2 cars. Finally, we have drag-car races. Drag cars are long and narrow and built close to the ground, designed for high acceleration.

The First Car Races

The first car races took place in southern England shortly after 1900. England led the field for several years, but France, Germany, and Italy soon got into the act. Within a few years, larger races were organized, until finally the sport became international. In the late 1930s a world championship was held, and soon there was considerable competition among the major auto manufacturers. Mercedes, Ferrari, Lotus, and others got into the act. As racing techniques developed, the cars changed; heavier, larger cars were replaced by lighter, more streamlined ones.

The Grand Prix became a huge event in Europe, drawing record crowds. Each country held its own Grand Prix. The cars were formula 1 and formula 2 cars. Big names on the circuit were Graham Hill, Jackie Stewart, Stirling Moss, and Niki Lauda. In America, the equivalent was the Indianapolis 500 with its Indy car. Some of the biggest names were Mario Andretti, Bobby Unser, Al Unser, and A. J. Foyt. American races usually took place on an oval track, whereas in Europe many were held on public roadways as well. Speeds of 200 miles per hour were soon common.

Stock- and modified stock-car races became common in the United States after World War II as cars were made more powerful. Several local organizations such as NCSCC and NSCRA were created, but it was soon obvious that a national organization was needed.

In December 1947, Bill France met with leaders of

the major associations across the United States, and plans were drawn up for a national racing association. Called NASCAR (National Association for Stock Car Auto Racing), it was officially formed on February 21, 1948. The first races were in modified prewar cars. In 1949, however, Bill France established the "strictly stock" car races, in which no modifications were allowed.

Many of the early races were won by Red Byron, who became the first major racing champion in the United States. Racing soon took off. Ten years later, in 1959, the first Daytona 500 was held, and since then stock-car racing has never looked back. Many household names were created: Richard Petty, Cale Yarborough, Bobby Allison, Dale Earnhardt, Jeff Gordon, John Andretti, and Bobby Labonte, to mention only a few.

Drag racing began in Southern California when hot-rodders began racing across the dry lakebeds of the Mojave Desert. The first organized events were held in 1931. Dragsters usually race in pairs along a straight-away, with bursts lasting only a few seconds. The object in this case is high acceleration.

Racing Techniques—Tires

Physics is, of course, critical in any type of car racing. The forces on a car change continuously throughout a race. They depend on many factors, and it is important that the race driver understand these forces and know how to control them. We have already discussed many of the topics of importance to a race driver—namely, aerodynamics, braking, suspension systems, engine power—and I will discuss several of them again in relation to racing. One of the most important features of a racing car is the tires: they are the contact between the ground and the vehicle, and they are certainly important in giving the vehicle its acceleration and speed.

Two of the top priorities for a race driver are keeping his car at its limit, and managing weight shifts. In this section we'll be concerned with keeping the car at its limit; this is what separates good drivers from mediocre ones. Earlier I discussed the traction circle. An understanding of it is critical to the race driver. Inside this circle we have traction, and outside, slippage. The radius of the circle represents the car's adhesion. The larger the area of the circle, the greater the grip of the tires on the road. This, in turn, depends on the downward force on your tires, and that's why it's important to keep this force as high as possible.

The vertical line through the circle represents acceleration and braking, with the top half representing acceleration. The horizontal line is associated with turning, either to the right or left. It is the objective of all good drivers to stay as close to the line as possible without exceeding it. If you exceed it, your tires will start to slip and you may lose control.

Looking at the circle we see that maximum acceleration is achieved when there is no turning. Indeed, if you turn when you are accelerating to the maximum you will pass beyond the circle and slippage will occur. Similarly, maximum braking can only occur when there is no turning. Finally, maximum turning to the right or left can only occur if there is no acceleration or braking. These points are shown in figure 74 as a, b, c, and d. They are particularly important, and usually the easiest to maintain while you are racing. The parts in between, however, are another story. In this case you have some braking or acceleration along with turning, and it's more difficult to know when you're exactly on the line.

The traction circle shown on the left in figure 74 is for the ideal situation, and this isn't always the case. If the tires in the front of the car have a different traction than those on the back, the traction circle will look dif-

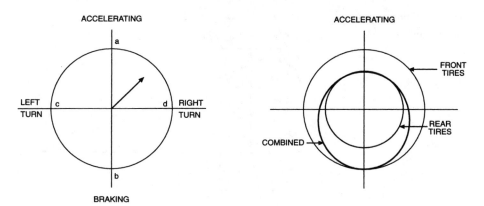

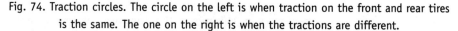

Fig. 74. Traction circles. The circle on the left is when traction on the front and rear tires is the same. The one on the right is when the tractions are different.

ferent. In particular, there will be a different traction circle for each set of tires, with one bigger than the other. The overall traction circle will then be oval shaped, as shown in the figure on the right. Cars with traction circles of this type are usually difficult to handle. They tend to oversteer by sliding the rear, and understeer during acceleration because of the low adhesion of the rear tires.

To get an idea of what happens when you go beyond the traction circle, we must consider the slip angle of the tires. Assume we are accelerating to the maximum and turn suddenly to the right. It's easy to see that we will pass beyond the circle. What happens if we do?

Let's begin with low-speed turns where the four slip angles are zero and assume the car is properly aligned. None of the tires is slipping and we are turning about the point shown in figure 75. To determine this point we draw lines perpendicular to the direction the tires are pointed. This gives the radius of the circle the car is turning around. In this case the weight of the car is distributed over the four tires.

Now assume that we accelerate or brake so that the weight distribution is not even. As a result the tires

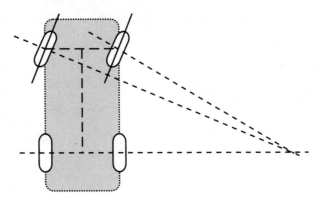

Fig. 75. Turning radius with no slip.

begin to slip. This means that they are not turning in the direction that the car is moving, and there is a slight angle between these directions—the slip angle. This happens, for example, when there isn't enough weight on the front tires. In this case we draw lines perpendicular to the slip angle of the front tires and notice where they intersect the lines from the back tires. We see that they intersect at a point farther out than in the above case. We are therefore now turning around a circle with a greater radius than expected. The result will be understeering, and the front end of the car will break away and begin to slide (fig. 76). For a race car driver in an understeering car, it is sometimes said that "the front end of the car goes through the fence first."

We can also have oversteering. This usually happens when the rear tires have insufficient weight on them and start to slip, giving the rear tires a slip angle. With both front and back tires having slip angles, as shown in figure 77, the radius of curvature of the circle we are moving around is less than expected. In this case the rear tires slip, creating oversteering, and for a race driver the car "hits the fence rear-first." In this case we also get slippage on the inside front tire. This car turns very sharply—much sharper than expected.

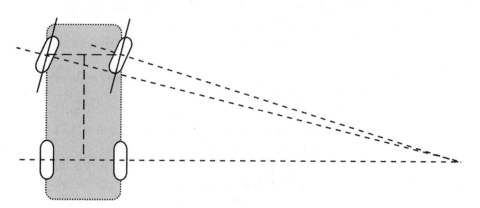

Fig. 76. Turning radius with slip angle in the front tires.

Weight in the Right Places

Weight distribution is also important to the driver. I'm referring here to the weight distribution when the car is stationary. Once it is moving the weight distribution changes, and that's a situation we have to look at separately. With the car stationary it's important to know how much weight is on each of the tires. This depends on the position of the center of gravity. The center of gravity is the point where, for all practical purposes, all of the mass of the car can be considered to be located.

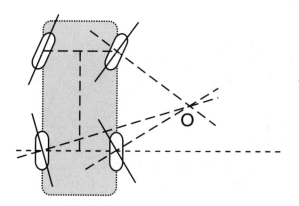

Fig. 77. Turning radius with slip angles in both the front and rear tires.

The force of inertia acts on all parts of the car, but we can simplify things by considering it to act only at the center of gravity.

How do we determine the position of the center of gravity? In practice it's usually quite difficult. The center of gravity is, in effect, the balance point of the car. It's important to remember, however, that it is the balance point in three dimensions. We can easily determine the balance point, or center of gravity, of a two-dimensional object by hanging it from two different points (fig. 78). For a three-dimensional object it's a little more difficult.

When a body is symmetric in shape and uniform in density, the center of gravity is usually easy to identify. For example, it is at the center of a sphere. If the density varies from point to point, as it does in a car, the problem becomes more complicated. And finally, if the object is irregular in shape, like a car, the problem can be very difficult. The center of gravity of a car can therefore usually be only approximated.

The main reason the center of gravity is so important is because cornering, acceleration, and braking forces can be considered to be acting at it. Roll and handling

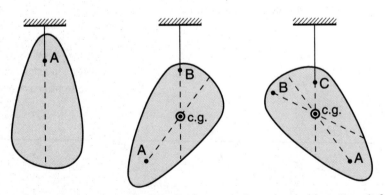

Fig. 78. Determination of center of gravity for a two-dimensional object. Hang it from several different points.

characteristics are therefore determined to a large degree by the position of the center of gravity.

Assuming that we have been able to determine the position of the center of gravity, we can now determine how the weight is distributed over the four tires. We will assume the weight on the two front tires is the same, and the weight on the back ones is also the same, so we will merely have to determine how much of the total weight acts on the front axle and how much acts on the back axle. The weight distribution is important because traction, steering, and handling depend on it. (See fig. 79.)

Assume that the distance between the tires is R, and the weight of the car, which acts at the center of gravity, is W. Also, assume the distance from the center of gravity to the front wheel is r_f and the same distance to the back wheel is r_b. The weight on the front axle is then

$$W(r_b/R),$$

and the weight on the back axle is

$$W(r_f/R).$$

Note that if more of the weight is on the front axle of the car, the grip of the tires in the front will be greater

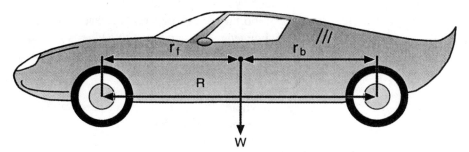

Fig. 79. Distribution of weight in the front and rear axes.

and the car will have better steering, but the braking may be poor. On the other hand, if more of the weight is on the rear of the car, the traction of the back tires will be better and it will have better braking and acceleration. Steering, however, may be a problem.

The exact position of the center of gravity—whether it is closer to the back or the front—therefore has an effect on the handling and turning characteristics of the car. The height of the center of gravity above the ground is also important. It is critical in relation to roll characteristics and weight transfer, as we will see in the next section.

It might not seem important, but the distribution of weight around the center of gravity is also crucial. A measure of this distribution is given by the moment of inertia. Like the center of gravity, the moment of inertia of a car is also difficult to determine. Furthermore, it is different for different axes throughout the car. For a symmetric object of uniform density, the moment of inertia is usually relatively easy to calculate, but when it is not symmetric it can be difficult.

In the case of rotational moment of inertia, the quantity mr^2 replaces the m of straight-line motion. As we saw in an earlier chapter, the force required to stop a body moving in a straight line depends only on the total mass of the object, even though the body is made up of numerous tiny subunits of mass. These subunits do not play an important role. This is not the case in rotational motion, however. The important role here is played by mr^2, so the distribution of mass is important. For an object such as a sphere, each little subunit of mass m at distance r gives a contribution. (See fig. 80.) For the body as a whole we have to sum over all m's and all r's. The moment of inertia is given by

$$I = \Sigma mr^2.$$

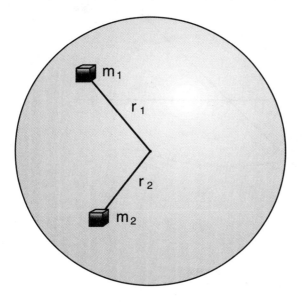

A car with a large amount of inertia will therefore have its mass distributed over a large volume. Its center of gravity may be at the same point as a similar body with a lower moment of inertia, but the car will handle differently. A car with a high moment of inertia will turn more sluggishly than one with a low moment of inertia, just as an object with high mass is harder to stop than one of low mass. This car will, however, be more stable than the car of low moment of inertia. A car of low moment of inertia, on the other hand, will turn easier and be more nimble, but it will be less stable than a car with a high moment of inertia.

The axis that the moment of inertia is taken around is also important. If it is taken along the long axis of the ellipsoid it is obvious that it will be less than if it were taken about an axis perpendicular to it (fig. 81). Remember that it is proportional to r^2, and the distance to most of the little volumes of mass will be greater in the second case.

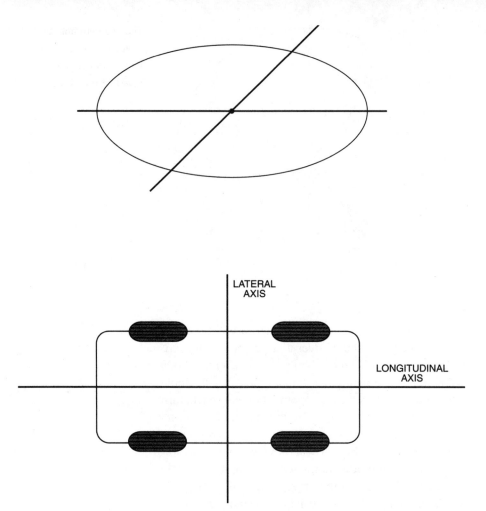

Fig. 81. Ellipsoid showing longitudinal and lateral axes (*top*); car showing longitudinal and lateral axes (*bottom*).

Because of the variation in density and its configuration, the moment of inertia of a car is difficult to determine exactly. At best we can usually get only an approximation. To a rather crude approximation, however, the shape of a car resembles the ellipsoid in figure

81, so we can determine the approximate moment of inertia around the longitudinal and lateral axes. The handling of the car will depend to some degree on the magnitude of these two numbers. There are obviously several cases we can consider. If our car had a large moment of inertia along its longitudinal axis and a small one around its lateral axis, it would react slowly when cornering, but it would be stable. To increase the cornering ability we would have to decrease the longitudinal moment of inertia. But we can't decrease it too far if we want it to remain stable. For the case where we have a large lateral moment of inertia and a low longitudinal one, we would have a car that is very responsive on turns, but not very stable in the longitudinal direction.

Balance and Weight Transfer

Balance is important both when a car is moving and when it is stationary. The first step is to make sure that the car is well balanced when it is stationary. If not, it is not likely to handle well. To a large degree, balance is associated with the suspension system of the car. As we saw earlier, a car has a suspension system for the front of the car and for the rear, and it is important that they work well together. First of all, the weight of the car should be distributed as equally as possible. If this is not the case, it is usually relatively easy to correct for it. Just adjust the stiffness of the springs. For example, if there is more weight on the rear, you will have to increase the stiffness of your front springs.

Balance when the car is in motion is a different matter as the weight of the car is continually shifting. The weight on each of the tires varies continuously during a race, depending on such things as acceleration, braking, and turning. It's important for a driver to know how the weight distribution of this vehicle has changed when he or she has performed a particular maneuver. Weight can shift in the up and down directions, to the right and left of the car, and to the front and rear. The up-down direction is different from the others because of gravity. The car can go all the way from being weightless when it is airborne, to weighing considerably more than it does when it is not moving. Earlier we talked about the effects of aerodynamics, and the downforce resulting from wings or the shape of the chassis. Both can considerably increase the car's overall weight.

In the other two directions the total weight is constant, and it can only shift around. Whatever weight comes off one tire or set of tires goes to another tire or set of tires. When the car is accelerating, for example, weight is shifted from the front wheels to the rear wheels. The total weight, however, remains the same. This is helpful if you need traction in the rear wheels, but it doesn't help the steering. With a decrease in weight on the front wheels you will tend to understeer. Similarly, if you brake, weight is shifted to the front wheels, and you will oversteer.

Suppose now that the car is rounding a corner. There will be a centripetal force acting toward the center of radius of the circle that the car is going around. In addition, there will be a horizontal force on each of the tires—a frictional force between the tire and the road. Furthermore, depending on the position of the center of gravity of the car, and its roll center, there will be a torque on the car. This results in the outside tires being loaded heavier than the inside ones. In other words, there is a shift of weight outward, onto the outer tires. If the torque associated with the shift is large enough, the car can roll.

One of the problems of unequal loading of the tires is that the overall grip of the tires on the track is reduced. It is therefore best to keep the loading as even as possible. Exactly what happens as a result of unequal loading depends to a large degree on the suspension system.

How much weight is shifted when we go around a curve? As it turns out, this is easy to calculate; the difference in weight on the outer tires as compared to the inner ones is given by

$$W_d = F_c\, h/R,$$

where F_c is the centripetal force, h is the height of the center of gravity above the track, and R is the distance between the tires, or wheelbase. The centripetal force can be calculated from

$$F_c = mv^2/r = mg(v^2/rg) = W \times \text{(centripetal accel. in g's)}.$$

W is the weight of the vehicle, and the centripetal acceleration is given by $a_c = v^2/r$.

Since we want the transfer to be as small as possible, h needs to be as small as possible and R as large as possible. In other words, the center of gravity should be very close to the ground, and the car as wide as possible. This is, of course, why race cars are built low and wide. It's important to note, however, that through F_c the transfer of weight also depends on the velocity of the car and the radius of the track.

To illustrate the above, consider a race car of 3000 pounds rounding a curve of radius 100 feet at a speed of 60 miles per hour. Assume the height of its center of gravity is 12 inches above the ground and the separation of its wheels is 100 inches. Substituting into the above formula shows that the centripetal force will be 7260 pounds, and the difference in weight on the inner and outer tires will be 872 pounds.

Earlier we dealt with the amount of weight transfer

to the front or rear of a car when the car is braking or accelerating. In this case the difference in weight is

$$W_d = Fh/R,$$

where F is now the inertial force obtained from

$$F = ma = mg(a/g) = W \times (\text{accel. in g's}).$$

Racing Strategy

One of the major questions for the race driver is: What is the quickest route around the track? The answer may not be the shortest route because you have to adjust your speed according to the forces on your car. If you are dealing with a curved path, the shortest distance will be the one with the smallest radius of curvature, but it is easy to see that you can't go around a curve with a small radius of curvature as fast as you can one with a larger radius of curvature. Earlier we saw that the maximum velocity for a curve of radius r is given by

$$v = 15/22 \ (ar)^{\frac{1}{2}}.$$

You want to keep the distance to a minimum, but at the same time keep your speed as high as possible so that the time is a minimum.

Let's begin with a right-angled turn, in other words, a turn of 90 degrees. We'll assume the track is flat and we'll have to consider the corner in isolation because, in practice, what comes after the corner usually affects how we go around it. For now, we'll assume that we are not concerned with what is around it, and merely want to get around it as fast as possible. (See fig. 82.)

The only two paths worth considering are a and b, shown as dotted lines in the figure. Curve a has the

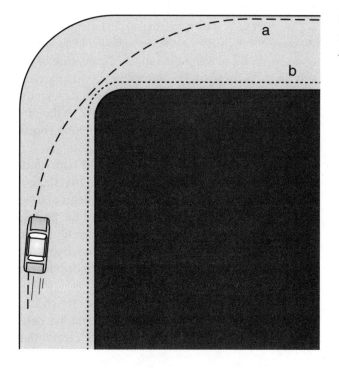

Fig. 82. Best strategy for rounding a right-angled curve.

largest possible radius of curvature, and curve *b* has the shortest. Curve *b* allows us to travel at a greater speed, but it's a longer path. So it's speed versus distance. Does the shorter distance of curve *b* compensate for the greater speed that we can maintain around curve *a*? A detailed analysis shows that it doesn't. It's easy to show that the quickest way around the corner is curve *a*, where we keep to the outer part of the track as we come into it, clip the apex as close as possible, then move to the outer part of the track after the turn. It is important to brake before the turn and keep the radius of curvature as constant as possible. Also, we shouldn't try to accelerate while going around the curve. These factors are, of course, all well known to experienced drivers.

The other type of turn we're likely to encounter is the 180-degree turn such as the one shown in figure 83. Again, we want to keep the radius of curvature of the path as large as possible. Therefore, as we approach this turn we should again take to the outside, clip the apex as closely as possible, and finally take to the outside again. As in the previous case, all braking should be done before we move into the corner.

As I mentioned earlier, we considered both cases above as isolated. Let's assume, however, that there is a long straight stretch after the above 180-degree curve. If this is the case the above strategy is not the best. In that case, we would want to enter the straight with as high a speed as possible. For this we will have to brake harder as we approach the curve and turn more sharply, as shown in figure 84.

In this case we will enter the straight stretch at a higher speed than we would have in the earlier case. But problems can result. We'll have to be careful that

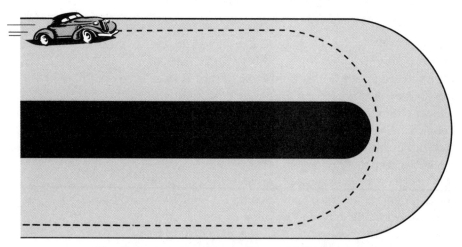

Fig. 83. Best strategy for rounding a 180-degree curve.

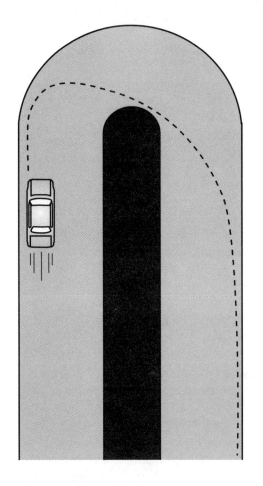

we don't strike the outside barrier, or go off the track, and with the extensive braking required as we come into the curve, other drivers may try to pass us.

Of course, all of the above cases are idealistic. In practice, they may be difficult to bring off. Other cars get in the way.

Another case worth considering is the banked curve. It's a little more complicated. In the case of a flat track

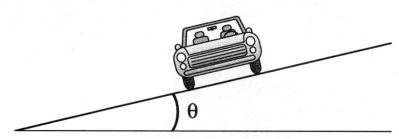

Fig. 85. Banked curve.

the centripetal force F_c must be compensated for by friction. If we don't want to rely on friction we can bank the curve as shown in figure 85.

It is easy to show that the proper banking angle for a velocity v is given by

$$\tan\theta = v^2/Rg.$$

And it's obvious from this that no one angle is right for all velocities. Curves are therefore banked according to the average speed of vehicles passing over them.

There are a lot of things to consider when racing: balance, weight shift, traction of the tires, and the best way to go around the track. It's a complicated process, but that's what makes racing so interesting.

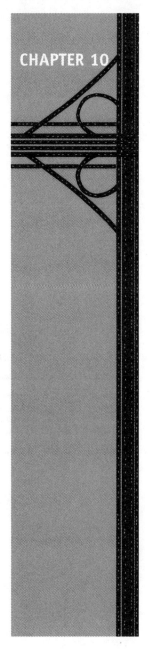

CHAPTER 10

Rush Hour

Traffic and Chaos

This chapter is going to be a little different from the previous ones: we won't be dealing with parts of a car. Nevertheless, the topic is an important problem in relation to cars. What I'm referring to is traffic, or more specifically, traffic congestion. Most of us have been tied up in a traffic jam at one time or another. This typically happens on Memorial Day, Labor Day, and on other holidays when the traffic is particularly heavy. Highways are built for a certain amount of traffic, and when it's exceeded there is congestion. No city is immune to it.

Things have gotten so bad in recent years in most of the major cities in the nation that several studies now under way hope to determine how to relieve the problem. Studies have, of course, been done before, but today we have tools that are much more powerful than in the past. One of the major tools is the computer. It is a necessity in such studies, but by itself is limited; realistic models of traffic and traffic congestion are also needed. And while a number of breakthroughs have been made in the area, there is still controversy.

In dealing with traffic we first have to look at the variables that determine it: flow, speed, and density. Flow and density are related, as flow is the number of vehicles passing a point per given time, and density is the number of cars in a given distance, say, a mile. Speed is also connected to them, since the more space cars have to themselves, the less they will interfere with one another and the faster they can travel. (See fig. 86.)

Several important computer studies of traffic have been done in both Europe and the United States. Sur-

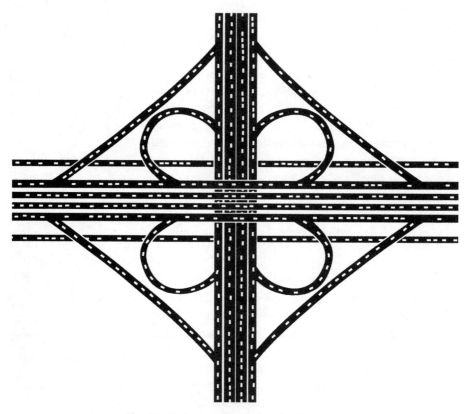

Fig. 86. Modern traffic. A cloverleaf system.

prisingly, many of them have come not from civil engineers, as you might expect, but from physicists. Civil engineers build roads, and the traffic on them is certainly their concern. But traffic has to be modeled, and a particularly close analogy to the thousands of cars on a freeway is the molecules of a gas, and molecules are the concern of physicists.

One of the physicists working in the area of traffic control is Bernardo Huberman of the Xerox Research Center in Palo Alto, California. Using computer simulations, Huberman has obtained some interesting results. He found that as the density of cars increased, the overall speed decreased, which was, of course, expected. But strangely, under certain conditions he found that flow could increase with increased density. According to Huberman, if the passing lanes became congested, they eventually reach a state where no one could pass. When this happens, the traffic begins to move as a solid block, and if this occurs, the average speed can increase, which, in turn, increases the flow.

Such a state—solid block, or synchronous, flow—is ideal for several reasons. In this state, cars cannot accelerate or change lanes, and therefore accidents are much less likely to happen. Studies have shown that most accidents on freeways are associated with acceleration, fast braking, and passing. Huberman compares the change from free flow to synchronous flow to a phase change, like the one that occurs when water changes from liquid to ice. But he shows that this state, while ideal in many ways, is also precarious. Increased flow can occur up to a certain point, but beyond it, synchronous flow can lead to disaster.

Dirk Helbing and Boris Kerner of Stuttgart, Germany, have also been simulating traffic flow on a computer. Like Huberman, they considered cars to be analogous to the molecules of a gas, with corrections being made for braking when drivers get too close to one another (molecules obviously don't worry about hitting one another). They found that many of the phenomena that occur in the motion of gas molecules are also seen in traffic flow. For example, when a flowing gas encounters a bottleneck, the molecules compress and set up a shock wave that passes back through the molecules behind it. This also happens when bottlenecks occur in traffic; everyone puts on their brakes and a wave passes back through the cars. This is, of course, something we would expect and we don't need a computer to tell us it would happen. But Helbing and Kerner found much more. Like Huberman, they showed that traffic can undergo a sudden change, or transition, from free flow to synchronous flow. They also showed that this synchronous flow initially gives an increased efficiency in flow rate.

Helbing and Kerner then looked at what happens as the density of vehicles increased further. They wanted to find out how it would affect the flow rate. As expected, as the density increased, the flow rate continued to increase, but then a point was reached where a decrease in flow occurred with further increase in density. Furthermore, as they pushed the density even higher, the flow rate continued to decrease.

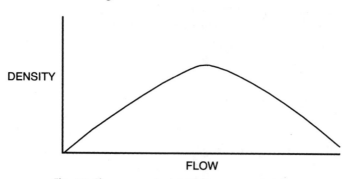

Fig. 87. Flow curve. A plot of flow versus density.

Plotting this up, we get a curve such as the one shown in figure 87. We see that there is a maximum on the curve, and beyond this point the flow efficiency drops off rapidly. This may not be surprising; after all, intuition tells us that something like this will eventually happen. But what Helbing and Kerner found next was surprising: under certain circumstances, the peak is not reached as the density increases. In essence, traffic could "tunnel" through the curve to the downward slope (fig. 88). And what was particularly significant was that it didn't take a major bottleneck, or even a moderate one, to cause this effect. Rather, insignificant changes within the flow could cause it, and when it happened, major traffic jams sometimes occurred that lasted for hours.

This phenomenon is reminiscent of what happens in chaos. Chaos is a state in which no organization whatsoever exists; in short, a state in which there is no order. We know that chaos develops in many situations similar to traffic flow. Weather is one of the best examples; it shows very sensitive dependence on initial conditions. A tiny change in pressure, or some other weather variable, at some point on Earth can create havoc a few days or weeks later, thousands of miles away. This is,

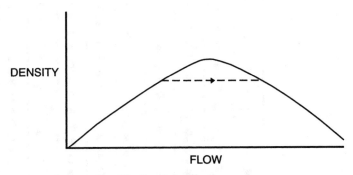

Fig. 88. How tunneling can occur.

indeed, why we can't predict the weather to a high degree of accuracy, despite a worldwide network of satellites and large arrays of supercomputers. Because of this sensitivity, chaos is defined as sensitive dependence on initial conditions.

Is it possible that traffic can become chaotic? Chaos can occur in gas molecules under the proper circumstances; we can, in fact, observe it in certain types of experiments. Helbing and Kerner's simulations seem to indicate that it is possible, but there is considerable controversy, and not everyone believes their results. (See fig. 89.) In many ways, the situation is similar to one that occurred several years ago. For decades, traffic engineers assumed that if there were problems (e.g., traffic jams and bottlenecks) in a particular section of a free-

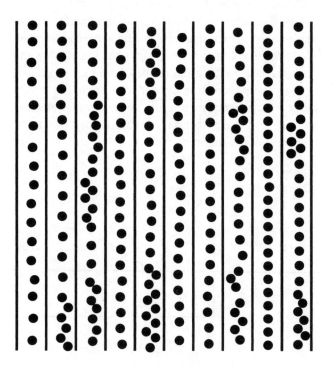

Fig. 89. Traffic in several lanes. Note that there is a "bunching up" in places just as there would be with molecules.

way network, it could easily be overcome by adding a few more lanes. In 1968, however, Dietrich Braess of Germany showed that this isn't necessarily true. His calculations showed conclusively that under certain circumstances, building a road or more lanes to alleviate a traffic problem can backfire. Indeed, it can actually decrease the traffic capacity. His result is now known as the Braess paradox.

Getting back to our question about whether chaos can occur in traffic, we have to begin with chaos itself. What exactly is chaos?

Chaos—A Brief Overview

Chaos was discovered in 1961. Actually, it was encountered earlier by Henri Poincaré of France, but he didn't see the significance of what he had discovered. He encountered it in the late 1880s in a problem related to the planets, but saw that it would involve a large number of tedious calculations and didn't pursue it. The person we therefore have to give credit to as the discoverer is Edmund Lorenz, an MIT meteorologist. It might seem strange that a meteorologist would make a major discovery such as chaos. But weather, as we saw earlier, is a place where chaos is prevalent.

To see chaos clearly, you need a computer. This is why Poincaré didn't follow up on it; he saw that the calculations were going to be long and complicated, and he didn't have a computer. But Lorenz did. He had written a computer program to check on certain types of weather patterns. One day in 1961 he decided to re-check a sequence he had just completed. He didn't need the entire sequence, so he started in the middle of it. As he began entering the numbers from the previous output he decided he didn't need all six digits that the computer had put out. To save time, he entered only the first three (i.e., .309 of .309547). He was sure it wouldn't cause a problem. After all, the fourth, fifth, and sixth decimal places were almost impossible to measure.

When he returned later to look at the new output he got a surprise. At first the results were the same, or at least very close, but they soon began to differ from the original sequence, and within a short time they were completely different. He couldn't believe his eyes. He tried it again, but got the same result. (See fig. 90.)

What his equations were showing was a very sensitive dependence on initial conditions. Checking further he found that even with a different system of equations he encountered the same problem. For him the surprise, however, was that if there was such sensitive dependence on initial conditions, there was no way you could make long-term predictions of the weather. We now know that five to six days is the limit for accurate prediction.

Looking into the details, Lorenz found that his results followed a double spiral. But strangely, the path never repeated itself. The loops stayed within certain bounds, but they never overlapped; they were random, or chaotic. This object is now referred to as a Lorenz attractor (figs. 91, 92). It looks like a butterfly, and because of its sensitive dependence on initial conditions it has led to the popular prediction: "The flap of a butterfly's wings in one part of the world can create a hurricane in another part a month or so later." Of course, this is an exaggeration, but it gives you some idea of the sensitivity.

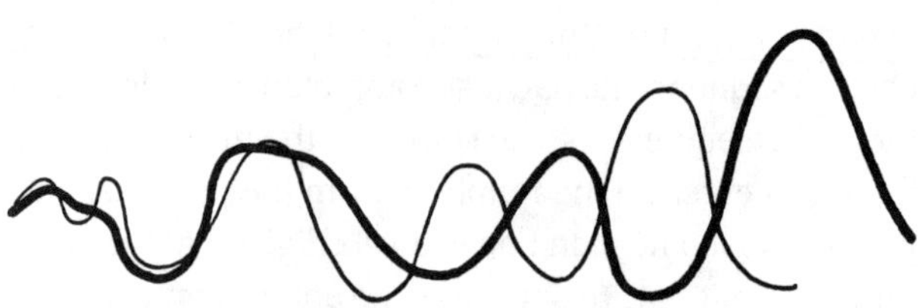

Fig. 90. Dark line is Edmund Lorenz's original line. Light line shows increasing deviation and chaos in a later run of his data.

The Isaac Newton School of Driving

Fig. 91. The Lorenz
attractor, *zx* projection
(Michael Collier).

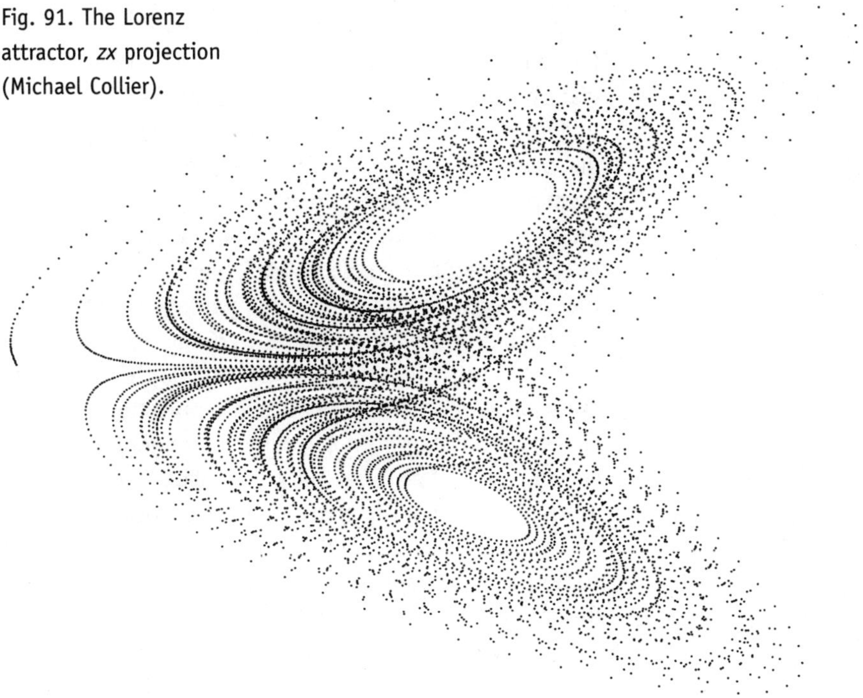

Fig. 92. The Lorenz attractor,
xy projection—a different
view (Michael Collier).

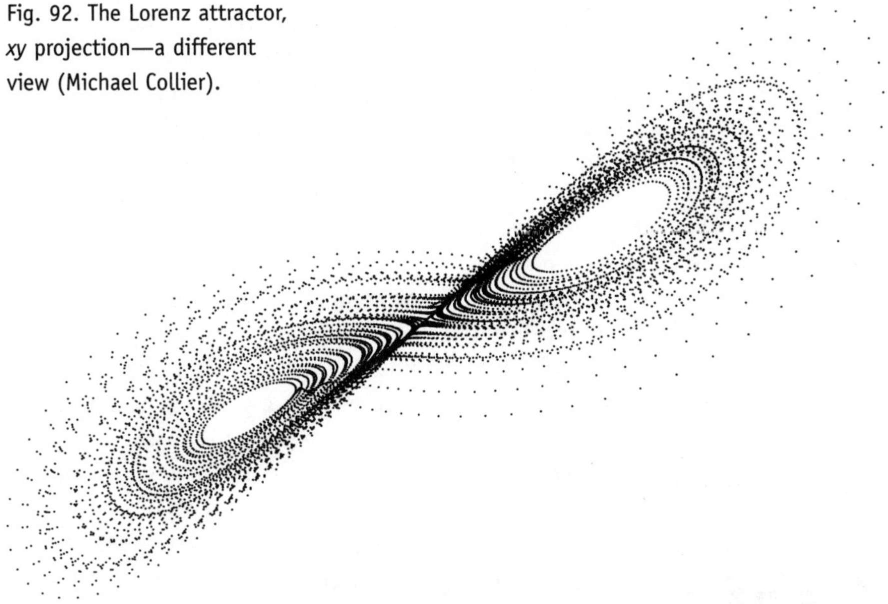

We now refer to this sensitive dependence as chaos. It is of tremendous interest to physicists and mathematicians because it has brought about a major change in the way they are forced to think about nature. Prior to this discovery, it was believed that the world was, for the most part, deterministic. In other words, with the appropriate equations we could calculate anything, particularly if we had a large computer. Chaos theory shows us that this isn't true. There are many things in nature that we will never be able to describe completely.

This is all well and good, but for us the important question is: Can traffic, or traffic congestion, become chaotic, and therefore strongly dependent on initial conditions? Is it possible, as in the case of weather, that a small perturbation in traffic flow can have a large effect somewhere down the line? This would mean that someone who applied the brakes suddenly when the traffic was particularly heavy could create a serious traffic jam. Helbing and Kerner found this to be the case in their simulations.

But let's look further. As it turns out there is more to chaos than Lorenz saw. Amazingly, there is order within chaos. To see it, we have to look at the work of biologist Robert May on the growth of biological populations, such as gophers, rabbits, and coyotes. The standard mathematical equation for such a population is

$$P_{\text{next year}} = RP_{\text{this year}}\,(1 - P_{\text{this year}}),$$

where P stands for population; it is a number between 0 and 1, where 1 represents maximum population and 0 represents minimum, and R is the growth rate.

May noticed something strange when he substituted numbers into this equation. When R passed 3, the line representing next year's population splits in two, indicating two different populations. One of the values was for one year, the other for the following year (fig. 93). The doubling, or splitting, is referred to as *bifurcation*. Increasing R further, May found that the output split again. Continuing, he found that bifurcations came faster and faster, and finally chaos appeared. Once there was chaos it was impossible to predict the behavior of a given population. (See fig. 94.)

If you look closely at the plot of the bifurcations in figure 94, you will see that the enclosed white regions continue indefinitely. They get smaller and smaller, but, upon magnification, they're always there. This is referred to as *self-similarity*, and it is an important property of chaos. Benoit Mandelbrot of IBM showed that this self-similarity is relatively common and occurs in many places. A good example is a map of a coastline. As you look closer and closer at the coastline, it con-

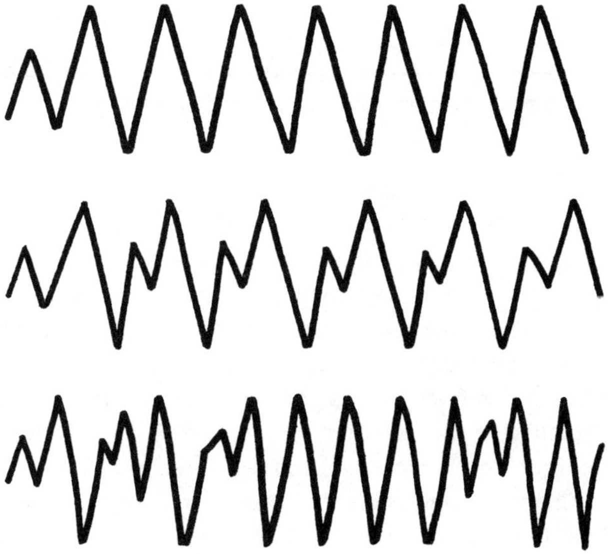

Fig. 93. Normal population growth (*top*); bifurcation to two populations (*middle*); chaos (*bottom*).

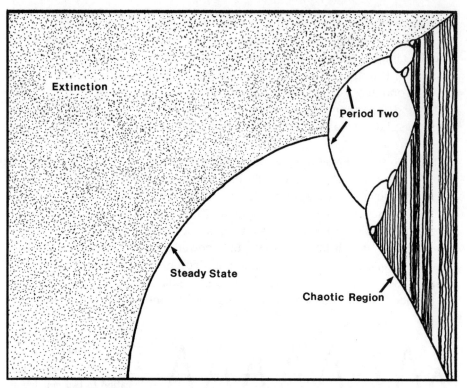

Fig. 94. Robert May's plot showing bifurcation and self-similarity.

tinues to show the same irregular structure on a smaller and smaller scale. Mandelbrot used the simple equation $z = z^2 + c$ to produce a particularly interesting self-similar image, where z is a complex number and c is a constant. (To get the image, the output of the equation has to be repeatedly substituted back into it.) Such structures are referred to as *fractals*. Fractal structure is exhibited by many structures in nature such as tree limbs, the body's arteries, and veins. (See fig. 95.)

Chaos has now been shown to be exceedingly prevalent. It occurs in a variety of areas, including weather

Fig. 95. Benoit Mandelbrot's famous figure showing self-similarity (George Irwin).

changes, population changes, disease epidemics, the stock market, public opinion, chemical reactions, and in astronomy.

Chaos and Traffic

Now, let's return to our question: Does chaos appear in traffic patterns? As we saw, Helbing and Kerner believe it does, and they showed that small fluctuations in traffic density can create traffic jams. But if chaos does play an important role in traffic, engineers will have to change the way they think about traffic control. It means that even when vehicle densities are well below what a highway is designed for, traffic jams can occur spontaneously, and there is little we can do about it. Widening roads or limiting on-ramp flow may not be the answer.

Many traffic engineers, however, do not take the results of the two German physicists seriously. They are not convinced that chaos plays an important role. To counter the skepticism, Helbing and Kerner tested their results by monitoring several German and Dutch highways, and they stated that their results were borne out. But even these comparisons have been questioned. Most people studying traffic agree that large traffic jams can occur for what seems to be no apparent reason, but they caution that the problem may be that no one has looked into the causes in enough detail. They suggest that such things as road conditions, animals running across the road, somebody suddenly braking to look at something, and so on could be at fault, rather than chaos. But if chaos is not at the root of many of the problems, what is? Let's look at another approach.

Complexity—A Brief Overview

Another approach to traffic congestion, which doesn't lead to such disastrous results, is through complexity theory. What is complexity? It's associated with chaos, but the effects are not quite as dramatic. Nevertheless, it also shows that there are problems in dealing with traffic congestion.

A good definition of complexity is hard to come by, but most scientists consider it to be "the edge of chaos." In other words, it is a complex phenomenon that has not yet become chaotic (and presumably will not become chaotic). Like chaos, it occurs in many areas, including earthquakes, the stock market, human brain waves, and the pumping of the heart.

The area of complexity we are interested in is called

cellular automata. It may seem at first glance that it has little to do with traffic, but as we will see, it does, and it has been used in traffic simulations. Cellular automata were invented in the late 1940s by physicist John von Neumann. He was interested in how a "machine" could be self-replicating. To him it was a mathematical problem, and his machine was a mathematical concept, but we frequently think of it as a living being.

According to von Neumann, a cellular automata is determined by four things:

1. A "space," which we can take to be a large board such as a checkerboard.
2. The number of states per cell (a cell is one square on the checkerboard).
3. The neighborhood of a cell.
4. A set of rules.

One of the simplest versions of von Neumann's scheme was invented by John Conway of Princeton University and is referred to as the "game of life." Conway's automata have only two states—black and white—and there are only three rules to his game:

1. A cell that is white at one instant becomes black at the next if it has three black neighbors.
2. A cell that is black at one instant becomes white at the next if it has four or more black neighbors.
3. A cell that is black at one instant becomes white at the next if it has one or no black neighbors.

In all other cases the cells retain their color. In Conway's game there are eight neighbors—four horizontal and vertical ones, and four in the diagonal directions. The idea of the game is to start with a certain figure, say, four black cells in a square, and see what becomes of it when the rules are applied. It's relatively easy to figure out what happens to a simple figure for a couple of moves, but for more complex figures and several moves, a computer is needed.

One of the goals of the game is to see whether a given configuration will disappear within a few moves or continue to exist in some form or other indefinitely. A particularly interesting starting figure is one that looks like a glider. It moves diagonally across the board. In fact, a simple glider gun can be set up that shoots out a stream of gliders (fig. 96).

What is especially interesting about the game is that it is virtually impossible to predict what will happen to a given configuration without carrying out the rules. If you try a few examples, it soon becomes obvious that these simple rules can give quite complex results.

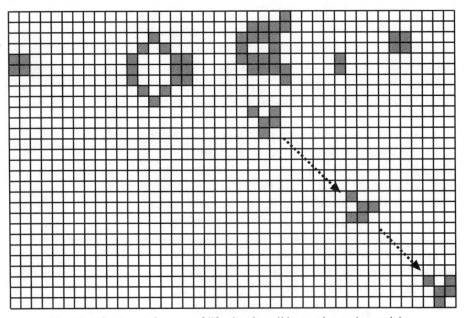

Fig. 96. John Conway's game of life showing glider moving to lower right.

It's important to point out that just as self-similarity plays a significant role in chaos, self-organization plays a significant role in complexity. That is, the system appears to organize itself after a certain number of generations. In other words, it manifests a coherent behavior, even though there is no central controller.

Complexity and Traffic

Like chaos, complexity has also been applied to traffic. Indeed, many traffic researchers regard traffic as a "self-organized" system. Traffic is, in many ways, like a biological system, and in this respect cellular automata, and Conway's "game of life," can be applied to it. In fact, the same type of rules can be used; in other words, neighboring cars can affect what happens to a given car. A system of this type has been set up on the computer by Christopher Barrett of Los Alamos National Laboratory. His model is called Transportation Analysis Simulator System, or TRANSIMS for short. Several well-known complexity theorists have stated that they believe the representation is a good one and that the application of the technique to traffic is valid.

One goal of such a program is to find out what the effect would be of adding new highway lanes or building new on- and off-ramps. Both of these can be simulated on the computer at much less cost than building a test model. TRANSIMS has been used to model the traffic patterns around such American cities as Albuquerque, Dallas, Fort Worth, and Portland, Oregon.

Not all traffic simulations are based on either chaos or complexity, however. A computer simulation at MIT called Intelligent Transport System Program mathematically takes into account the habits of drivers—factors such as speeding, frequently changing lanes, and so on. It has been used in testing various road-planning scenarios around a number of large cities and has been found to be very effective.

Computer programs will no doubt do a lot to help in the future and, as I mentioned earlier, simulations are much less costly than actually building roads. Ultimately, however, traffic will have to be controlled in a more direct manner with such projects as automated highways. Let's turn, then, to automated highways.

Automated Highways

Computer studies of traffic patterns and problems are certainly essential, and they will no doubt help guide us, but traffic is always going to be a problem, and with our increasing population it's not likely to ease up soon. The most obvious way around the difficulties, at first glance, is to build more highways. But they are expensive.

At the present time a typical freeway lane can handle about two thousand vehicles per hour. This number could be increased significantly by using a computer-controlled automated vehicle system. In other words, if we controlled the vehicles on the highway using computers within the vehicle, along with computers and various devices along the road, we could increase the efficiency significantly. In fact, it has been shown that we could increase the flow up to six thousand vehicles per hour—a threefold increase. And such a system would be much less expensive than building a new highway system capable of tripling the traffic flow.

The idea of self-driving vehicles has been around for a long time, and many tests have been made of the feasibility. But many of these tests are several years old, and in recent years electronic devices, lasers, wireless communication, and so on have advanced so much that the idea is now taking on a more serious tone. There are several possibilities for an automated system. Entire highways could be automated, or only single lanes. It is also possible that automated vehicles could mingle with the usual traffic. Either way, automated traffic would likely require special on- and off-ramps to the freeway. In an on-ramp, for example, the driver would switch from manual to automated driving. The computer would then sense the cars on the freeway and bring the automated car into the traffic safely. One technique might be that the driver programs a navigator within his vehicle according to his required destination. The navigator could then select the best route and take over.

Another possibility is a transition lane in which the driver switches the car from normal to automated driving. In this case the driver would have to enter the highway under normal control. Once his vehicle was in automated mode, it could mingle with the other traffic, changing lanes and so on, or it may stay in a special lane reserved for automated cars.

A number of devices would be needed within the vehicle for automated driving. A computer would be essential, but other gadgets such as video cameras, radio, or infrared lasers would also be needed to detect vehicles ahead, behind, and beside one's car. Devices would also have to be attached to the steering and braking systems to control them. Furthermore, digital radio equipment would be needed so that the automated vehicles could communicate with one another. Contact with the road itself could be maintained using magnets buried alongside the road; a magnetometer would be needed to detect them. Small computers would also likely be needed along the road to monitor the traffic.

The devices along the road and within the vehicle would share the responsibility for keeping cars separated by a safe distance to avoid accidents. Another possibility with automated vehicles is a "platoon" or series of vehicles that are tied together, much in the same way the coaches of a train are linked. There would, of course, be no physical link between them. This would be particularly effective with buses.

What about the passengers in an automated car? What would he or she do once the vehicle was in the automated mode? They would be relieved of any responsibility of driving so they could relax and even rest or sleep. The system would, of course, have to have a "wake-up" call once the destination was approached, so the driver could go back to normal driving.

It will no doubt be years before fully automated systems are available. One of the problems of such a system is cost. In all likelihood, such systems will be phased in, and some of the first steps will be advances in cruise-control systems. Brakes, for example, may be automatically applied when the vehicle gets too close to another one. Drowsiness is another problem, for many accidents are a result of people falling alseep at the wheel. Devices within the vehicle could sense when the driver is becoming drowsy and announce it over a loudspeaker.

"OnStar" and various navigational systems are already being used in cars, and they will no doubt become more sophisticated and advanced over the next few years.

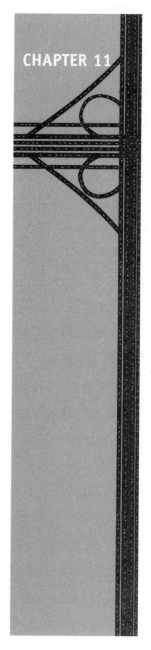

CHAPTER 11

The Road Ahead

Cars of the Future

After seeing the movie *2001: A Space Odyssey* many years ago, I left the theater wondering whether the world would really be like that in the year 2001. Well, 2001 has come and gone, and it doesn't seem as if fiction, or at least science fiction, has kept up with fact. There are, of course, many things around that we probably never dreamed about a few decades ago. A good example is the Internet. But our vision of streamlined cars silently hovering over a pad of cushioned air, opening on command and zooming off to our destination after a few quick words, is not quite here yet. A lot of work has been done on voice recognition lately, however, and our cars may soon open and start using voice commands. And navigators are now becoming common in luxury cars. So the day when you can just tell your car where to go may not be too far off.

At the present time, though, we have more immediate concerns. Our oil reserves are not unlimited (as much as we would like to think they are), and the day of reckoning may not be too far off. Furthermore, pollution from car exhausts is now considered to be one of the major problems around most cities. The estimates of how many deaths it has caused may not be an exaggeration. So, while the future cars we dreamed about a few years ago may still be a few years off, engineers are racking their brains looking for solutions to the problems of fuel economy and pollution.

At one time it was thought that the electric car would be the solution. With virtually no pollution and excellent mileage, it was going to solve most of our problems. All we would have to do is plug it into an outlet in our garage each night. But electric cars need batteries—particularly good ones if they are to store a lot of electrical power—and unfortunately battery technology hasn't kept up with our needs. For the present, our best bet seems to be hybrids, or HEVs (hybrid electric vehicles).

HEVs

Anything that uses energy from two different sources is called a *hybrid*. And HEVs use energy not only from electricity but also from an internal combustion engine, meaning we still haven't gotten rid of our gasoline engine. But with an auxiliary electric energy source, the internal combustion engine can be made much smaller and more efficient, so we're still ahead.

Hybrids aren't new. The first patent for a hybrid was actually taken out in 1905 by the American engineer H. Piper. His electric motor vehicle was augmented with a small gas engine. The first electric hybrids weren't built until 1912, however. Then everything went into a slump. Significant advances were made in internal combustion engines and everyone forgot about electric cars. There was a brief resurgence of interest in the mid-1970s with the gas crunch, but it was short lived. Then about the mid-1990s another round of interest began, and in 2000 the first two electric hybrids, the

Honda Insight and the Toyota Prius, were put on the market.

The efficiency of the hybrid is less than that of the purely electric car, and it is not environmentally friendly, nevertheless it is much better than gasoline-engine vehicles. In the internal combustion engine only about 20–25% of the energy in the gasoline is converted to useful energy. An electric motor, on the other hand, converts 90% of the energy from a storage cell to useful energy, but there are, of course, other energy losses. Nevertheless, on average, HEVs are at least twice as efficient as internal combustion engines.

An important breakthrough in electrical-powered cars came in the early 1990s. Until then virtually all electrical vehicles used DC motors, as they were considerably easier to run directly from a battery. But with the introduction of new technology, the DC motor was replaced by the AC motor, and it is now used in all electric vehicles. Its advantage in cars is that it is more efficient and reliable than the DC motor.

One of the advantages of HEVs is that they consume no energy when the vehicle is resting at a red light, nor when the vehicle is coasting. In addition, when it is slowing down—in other words, when the brakes are applied—energy can be redirected back into the battery. This is referred to as *regenerative braking,* and it is a particularly exciting aspect of HEVs.

Part of the HEV's efficiency comes from the fact that the gasoline engine in them can be made smaller than in conventional cars. The gasoline engine in conventional cars is large for one reason—so that you accelerate rapidly when you floor the pedal, and of course, so you can climb hills easily. In a nutshell, its huge horsepower is directed at its peak performance, but this peak performance is used less than 1% of the time. Most of the time the vehicle is rolling down the highway at a uniform speed, using only about 20 horsepower. The hybrid is designed for average driving conditions, so its gasoline engine can be much smaller and more efficient than conventional engines. The extra "spurt" that is occasionally needed comes only when it is needed; it's like an auxiliary power.

Three types of configurations are possible for hybrid vehicles: series, parallel, and dual mode. In the series configuration only the electric motor is connected to the wheels (fig. 97). The major function of the gasoline engine and generator is to keep the battery charged. It is usually charged from 60% to 80%. When the charge

Fig. 97. Series configuration for HEV.

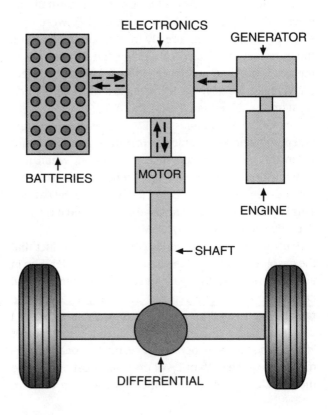

reaches 60% the electronics in the vehicle start the gas engine, which in turn runs the generator, recharging the batteries. When the charge in the batteries reaches 80% the engine is turned off. Incidentally, all the power to the back wheels is supplied by the electric motor.

The main alternative to the series configuration is the parallel configuration, in which both the electric motor and the internal combustion engine are capable of turning the back wheels (fig. 98). This system allows the vehicle to accelerate faster, but it is generally not as efficient. In this case the electric motor assists the gasoline engine during startup and during acceleration, or when there is a heavy load. It's important to note that

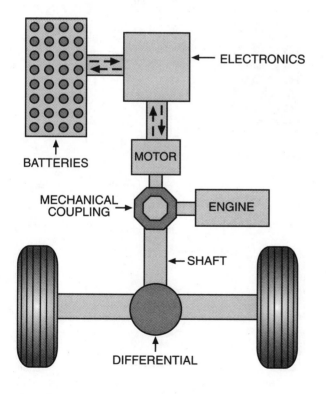

Fig. 98. Parallel configuration for HEV.

 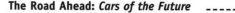

both the electric motor and the gasoline engine can turn the transmission at the same time. The transmission, in turn, powers the back wheels. Advanced electronics allows the motor to be used as both a motor and a generator (a motor used in reverse), therefore a generator is generally not needed in this type of vehicle.

The third type, the duel mode hybrid, is basically a parallel hybrid with a generator for recharging the battery (fig. 99). During normal driving, the engine powers both the back wheels and the generator. The generator, in turn, supplies power to the batteries and the electric motor.

Fig. 99. Dual mode configuration for HEV.

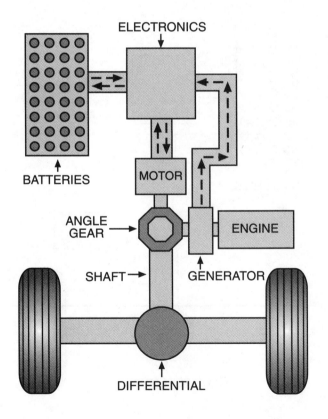

The Honda Insight uses a slight modification of the above configuration: it has an electric motor coupled to the engine. This "auxiliary motor" assists the gasoline engine, providing extra power when the vehicle is accelerating or going up a hill. It also starts the engine, so no starter is needed, and finally it provides some regenerative braking for capturing energy during braking. The electric motor is not sufficiently powerful to run the car by itself, and its major role is assisting the gasoline engine.

The gasoline engine in the Insight is three-cylinder and weighs only 124 pounds. It produces 67 horsepower at 5700 rpm, but this is sufficient to take the car from 0 to 60 in 11 seconds. When the electric motor is assisting the gasoline engine, the total horsepower is 73 at 5700 rpm. The difference is only 6 horsepower, but the torque output is increased significantly. It is 66 lb-ft at 4800 rpms without electric assist, and 91 lb-ft at 2000 rpm with it, which is a significant increase. It uses a conventional five-speed manual transmission.

The Toyota Prius is quite different from the Honda Insight. The Prius uses the dual mode configuration, and unlike the Insight, its electric motor is capable of moving it. In fact, the electric motor takes it up to a speed of 15 mph before it is switched to the gasoline engine. The gasoline engine of the Prius produces 70 horsepower at 4500 rpm, and the electric motor produces 44 horsepower at 1000–5600 rpm.

A unique feature of the Prius is its transmission: it is a power-splitting device. The gasoline engine, generator, and electric motor are hooked together through it, and because of this, the electric motor can power the car by itself, or the electric motor and gasoline engine can power the car together. The power split device is a planetary gear set with the electric motor connected to the ring gear of the gear set. The generator is connected to the sun gear, and the engine to the planetary carrier. All these have to work together and the output depends on this.

Space-Age Technology: Fuel Cells

The electric hybrid has much going for it, but it still has to compete with conventional cars (at least for the next few years) if it is going to be a commercial success, so any improvements that can be made to it are welcome. One of the biggest disadvantages is the batteries, which are generally larger and not as fast as we would like them to be. Is there any way of getting around this? Indeed, there is. Space-age technology has supplied us with the fuel cell (fig. 100). Fuel cells have been used in some of the space flights and have a definite advantage over conventional batteries. Ordinary batteries, such as dry cells, store a fixed amount of energy, and have to be recharged when they run down. Fuel cells, on the other hand, run on fuel (as their name implies), and therefore, as long as fuel is available, they continue to run. The fuel that is needed is hydrogen, and it is plentiful.

Unfortunately, the fuel cells used in spacecraft were

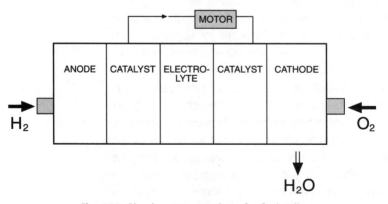

Fig. 100. Simple representation of a fuel cell.

generally too large and inefficient to be used in cars. In the 1980s, however, the Canadian engineer Geoffrey Ballard decided to see if he could make them smaller and more efficient. He was interested in using them in buses. After some experimentation he decided to try what is called "proton exchange membrane technology." In this case, hydrogen is introduced at the anode; from here it passes to a catalyst. The catalyst, which is usually platinum, breaks up the hydrogen into a proton and an electron. The free electrons are then siphoned off into an external circuit where they power an electric motor; they return to the catalyst near the cathode. Beyond the catalyst is a material that allows current to flow, referred to as an *electrolyte*. The protons pass through this electrolyte on their way to the cathode. Oxygen (or usually just air) is introduced at the cathode. The protons and oxygen come together to produce water, which is the exhaust. Thus, the only emission from the fuel cell is water, which makes it a particularly "clean" cell.

Individual fuel cells don't generate much electricity, but Ballard was able to "pancake" them into a stack that generated considerable power. As we saw earlier, the ideal fuel for such cells is pure hydrogen, but hydrogen is a problem. It is a highly volatile gas and is difficult to store. Considerable hydrogen would be needed by the cells so the containment vehicle would have to be relatively large, and it would have to be built secure. Furthermore, if such cells became common in cars, a whole new infrastructure would be needed—namely, hydrogen service stations.

Fortunately, there is an alternative, but it has a serious effect on the efficiency of the system. Hydrogen is available from many hydrocarbons such as methanol, ethanol, and gasoline, which are generally easier to store. The process of extracting it, however, gives off pollution. Still, the overall pollution is considerably less than that of conventional cars. A "fuel reformer" is needed for the extraction.

For those with a chemical bent, the chemical reactions in the cell are as follows:

At the anode: $$2H_2 \rightarrow 4H^+ + 4e^-$$

At the cathode: $$4e^- + 4H^+ + O_2 \rightarrow 2H_2O,$$

which gives overall $2H_2 + O_2 \rightarrow 2H_2O$.

The efficiency of the fuel cell is approximately 80%. This means that it converts 80% of the energy content of hydrogen over to electrical energy. With a fuel reformer and methanol, however, there is a drastic reduction—down to about 30% to 40%. The overall efficiency in this case is about 24% to 32% which is still better than gasoline's 20% efficiency.

The Flywheel

Another method of storing energy is through the use of a flywheel. The flywheel has been around a long time in cars. It is the heavy, toothed wheel that is mounted to the rear of the crankshaft, and its purpose is to smooth out the separate power surges imparted to the crankshaft as the various cylinders fire. But I will not talk about this particular flywheel. In a hybrid or electric car, a flywheel would serve to store energy, and in this sense it acts as a battery. Flywheels store energy mechanically as spinning, or angular, kinetic energy. An electric motor is used to accelerate the rotor up to speed; its energy can then be extracted as electrical energy using the motor as a generator. When energy is extracted from the rotating flywheel, it slows down, but it can be "recharged" by speeding it up again. The advantage of the flywheel is that a large amount of energy can be extracted in a relatively short period of time. The extraction rate is much quicker than that from a battery.

Flywheels are about 80% efficient. Their major energy loss is a result of friction, but they can be made almost friction free by use of magnets and a vacuum. One of their drawbacks is that for optimum efficiency, the flywheel needs to spin at the highest possible rate, and this can be exceedingly high—of the order of 60,000 rpm. This is a bit mind boggling if you think about it. It's hard to visualize something that is going around 60,000 times in a minute—that's a thousand revolutions every second.

The kinetic energy stored in a spinning flywheel is given by the formula

$$KE = \tfrac{1}{2} I \omega^2,$$

where I is the moment of inertia and ω is the angular velocity. The moment of inertia I depends on the shape and configuration of the flywheel. It is given by

$$I = kmr^2,$$

where k is an inertial constant, such that

solid cylinder	$k = \tfrac{1}{2}$
ring	$k = 1$
solid sphere	$k = 2/5,$

and m is mass.

The high rate of spin of a flywheel creates a serious problem. It gives rise to a centripetal force on the disk, which is given by the formula

$$F_c = mr\omega^2.$$

This formula shows that most materials cannot stand up to spin rates of the order of 60,000 rpm. They are ripped apart by the tremendous force. This means that the tensile strength of the material is especially important. A particular danger is a breakdown of the containment vessel; it has to be sufficiently strong to contain the debris. With such high angular velocities, the debris would be sent outward with considerable force and would be dangerous to anyone around it.

Before I leave the topic, just a note of caution in regard to the above formulas. In dealing with angular velocity you may have thought of it as revolutions per minute. I have used this unit several times, and if you have any old phonograph records, you no doubt remember that they rotated at either 78 rpm or 33⅓ rpm. Scientists, however, prefer to measure rotational velocity in terms of radians, where one revolution is 2π radians. One radian is therefore about 57 degrees. So if you're working on a problem involving one of the above formulas, it is important to convert the angular velocity to radians/minute (or second).

Getting back to flywheels, a particularly interesting test car was built in California by Rosen Motors using a flywheel. The company didn't supplement it with an internal combustion engine, as you might expect, but it used a turbogenerator. The turbogenerator was powered with gasoline, and it kept the flywheel spinning at the proper speed.

Ultracapacitors

Batteries and flywheels aren't the only places where electrical energy can be stored. In most electronic cir-

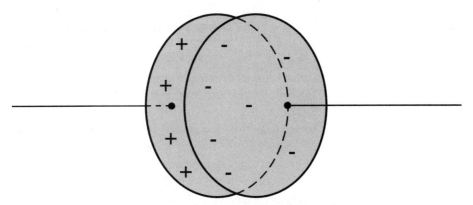

Fig. 101. Simple representation of a capacitor.

cuits there are dozens of capacitors, which also store electricity. But we usually don't think of them in terms of long-term storage. In fact, if you charged up an average run-of-the-mill capacitor and left it on the shelf, half of its charge would leak away in 24 hours or less. An alkaline battery, on the other hand, can sit on the shelf for three or four years and still retain 80% of its energy. So, as far as storage devices for energy in cars are concerned, the main problem with capacitors is leakage. But capacitors have many advantages over batteries. They can load and unload their energy much faster, which is important when a car accelerates or when it is going up a hill.

Let's look briefly at the capacitor (fig. 101). It is composed of two plates, with one plate positively charged and the other negatively charged. There is therefore a potential difference between the two plates. The greater the charge on the plates, the greater the potential difference. Assuming there is a vacuum between the plates, we can write the formula governing it as

$$q/V = C,$$

where q is the charge in coulombs and V is the potential difference in volts. C is capacitance. The units of capacitance are coulombs/volt, or farads. A farad, as it turns out, is a particularly large unit, so in practice, what we usually encounter is microfarads (a millionth of a farad). A dielectric is usually placed between the plates to increase the capacitance. The capacitance can also be increased by reducing the distance between the plates or increasing the area of the plates.

For completeness I'll include the formula for the energy of a capacitor. It is

$$E = \frac{1}{2} CV^2.$$

Within the last few years, supercapacitors, or ultracapacitors, as they are known, have been developed. They have millions of times the energy-storage capacity of traditional capacitors. Furthermore, they can unload their energy quickly. Indeed, they can unload their energy ten to one hundred times as fast as batteries. So far, though, they can't store as much energy as a battery. The total energy of an ultracapacitor is about 10% that of a lead acid battery of the same weight. But compared to ordinary capacitors, they leak charge at only about one-fiftieth the rate.

Ultracapacitors are obviously ideal for use in cars and electric hybrids. And they have been used in GM's new electric buses in New York and are scheduled for several different cars produced in the early 2000s.

HCCI

I have many fond memories of my first car. Some of them seem humorous now, but at the time they didn't seem funny to me. One of them was the stubbornness of the engine. I would arrive at my destination, turn the key off, and jump out of the car. But the car would keep running. I'd get back into it and bang on the dashboard a few times, but it wouldn't help. After a couple of minutes, though, the engine would stop. I thought it was a little crazy and had no idea what was going on. A mechanic told me it was "after run" and not to worry about it. Nevertheless, to me it was a bit of a nuisance. Surprisingly, engineers are now taking a serious interest in it, and it is now referred to by the impressive name "homogeneous-charge compression ignition combustion," or HCCI for short.

Given that the spark plug is no longer firing, the process has to be compressed heating—the same process that powers the diesel. There is, however, a distinct difference between it and the diesel. In the diesel engine, fuel is sprayed into the cylinder during the piston's compression stroke, but the turbulence of the flow allows only partial mixing. It is therefore referred to as a heterogeneous process. In the HCCI process, the mixing is thorough and the combustion temperature is usually lower. Furthermore, the amount of burning fuel is low compared to the volume of air. Therefore, the engine produces few pollutants, making it a relatively clean process.

Although there has been a lot of interest in HCCI lately, there are problems. It works well when the engine is under no load, but when a load is added the engine tends to slow down, and with increased load it knocks badly. One of the main problems is that because the fuel is thoroughly premixed with air, the mixture ignites all at once, unlike diesel fuel, which has a more extended combustion. Adjustments can be made for this, but they give rise to increased pollution. Therefore, it seems that HCCI may be suitable only for light loads, and might be better as an assist, rather than as a major engine. For example, it could be used in dual-mode engines. In these engines at high loads, the usual spark plug ignition is used, and at lower loads HCCI takes over. Considerable work is now going on in relation to using it in a hybrid vehicle.

The Compressed-Air Car

What we would really like is a car that runs on something that is really abundant, and therefore cheap. A car that runs on air would be a good example. After all, air is all around us, and any way you look at it, it should be cheap. Well, work is indeed under way on a compressed air engine. Most of the work has been done in France. The new engine will have zero pollution, but of course we need energy to compress the air that goes into the tank, so there will be pollution somewhere (i.e., in an electric plant).

Compressed air is stored in a tank, just as gasoline is stored in the conventional vehicle. The pressure has to be of the order of 4400 psi (pounds per square inch). The air from this tank is fed to the engine, where it is used to push the pistons down. The pistons, in turn, move the crankshaft, which powers the wheels. Like the electric vehicle, the compressed air car is a hybrid. The engines made in France can run on either compressed air or act as internal combustion engines. At low speed, below about 40 miles per hour, it runs on compressed air, but above it, it uses gasoline. The compressed air tanks are filled using household electric power, but a rapid three-minute recharge is possible using a high-pressure air pump.

A variation of the compressed air-powered engine is the *cryogenic heat engine.* Liquid nitrogen is used as the propellant in this case. Nitrogen makes up about 78% of our atmosphere, so it's obviously abundant. The liquid nitrogen is stored at –320° F. It is vaporized by a heat exchanger; the nitrogen gas that is forced into the heat exchanger expands by a factor of 700, and it is this expansion that powers the vehicle. The expanding gas pushes on the engine's pistons, just as the expanding compressed air did. Again, little pollution is generated, but as in the case of the compressed-air car, electricity is needed to get nitrogen out of the air.

Only time will tell if either of these versions will power the car of the future.

Telematics

New power sources for older cars and new types of cars won't be the only automative innovations of the future. Telematics, which deals with wireless communications in cars, is also something that is going to change our lives. We already have "OnStar" but it's merely a preview of things to come. In the car of the future you will press a button on your dash and say "gas." Moments later a voice will come on describing the position of the six nearest gas stations along with their brand of gas and the price. You press the button again and say "motel." Again the voice comes on and gives you a list of the nearest motels. All this is possible because of a Global Positioning System (GPS) locator. The signal from your car is relayed to a satellite miles away. It locates your car, then, after interacting with the Internet or some other electronic system, it signals back to you with its list.

Cars of the future will be equipped with a wireless transmitter and receiver, an antenna, text-to-speech capability, voice recognition, and a GPS unit. They constitute the main components of a telematic system. The major functions of the telematic system will be safety and security services. If your car breaks down, assistance will be sent immediately. Such a system will be extremely helpful in tracking stolen vehicles and may help decrease car theft. Other functions of which the system may be capable are remote door opening and an automatic 911 call when the airbag is deployed.

Navigation systems are already being used in some of the more expensive cars, and they will no doubt become more and more sophisticated. With the mere announcement of your destination, several routes will appear on the navigation screen. Furthermore, the system may have the ability to check the traffic on each of the routes and tell you which is the best route. Expected weather conditions during your trip may also be accessible. In fact, you may be able to make reservations or buy tickets to an event as you travel.

Voice recognition is going to be critical to this new revolution. Considerable work is now going on in this area, so it's not likely to be far off. The way a person says his or her name (or any other word) is as characteristic of the person as his or her fingerprint. Your voice box creates a sound wave that is made up of various frequencies and amplitudes. It's unlikely that anyone else could say your name with the same distribution of frequencies and amplitudes that you give to the sound. You're no doubt familiar with this in relation to the telephone. You answer and know immediately who is on the other end from the sound of the person's voice.

Voice recognition is particularly important because hands-on devices within the car create a distraction for the driver. For the most part, voice commands or an answering voice do not. The problems with cellular phones are well known, so telematics will no doubt try to stay away from anything that distracts the driver.

It is estimated that by 2006, about 50% of all new cars on the market will have telematic systems.

Other Devices

Telematics may very well be considered *the* big change over the next few years, but a lot of other high-tech gizmos are likely to appear. I will note some of them. "Night vision" is already available in some cars and is likely to become more common soon. In this case, infrared beams project farther out than the driver can see. The technique has been used in the army for years, where night goggles allow soldiers to see through the darkness. Infrared radiation is actually "heat" radiation, and we have very sensitive heat-detecting devices that work just as well in darkness. An image of what is beyond the driver's vision is projected on a screen, a feature that would be particularly valuable in relation to animals running across the road at night.

Tire-pressure monitors are also likely to appear soon. They will tell the driver what the pressure in each of the tires is, and alert him or her if any of them are low. Small TV cameras within the car for monitoring children in the back seat may soon be an option as well. Custom air-conditioning is also coming. There is not a lot of physics in all of this, but most of the devices do rely on physics in one way or the other.

The Far Future

We've seen that not too far in the future, cars may drive themselves using onboard computers and sensors in the road. Along with this prospect, however, will likely come another important advance: cybernetics. Cybernetics is the use of intelligent machines to amplify human capabilities. A device within the car reads your brain waves and understands your intent. And, assuming a certain intent, it takes over the car and performs the tasks better than you could have done. We already have crude forms of cybernetic control in cars. One of the best examples is ABS brakes. The ABS system senses that the car is sliding when you apply the brakes. Normally you would begin pumping the brakes to stop the slide and slow the car down, but with ABS the car takes over this task, and it does it much faster and better than you could. ASR, or acceleration slip reduction, and traction control (TC) are other examples. The system senses that one tire is spinning too fast, and shifts the load to the other tires.

How would the cybernetic system work? It would have to read your brain waves and make a decision based on them. In other words, it would have to determine what you wanted by analyzing these waves. Since everybody's brain waves are slightly different, just as our voice waves are, the system would have to be calibrated according to your particular waves. To some degree, fighter pilots, who occasionally have an overload of things to do in the cockpit and can't react fast enough, are already being helped by this system.

Cybernetic control would obviously be invaluable under certain circumstances. One case would be just before an accident. The system could sense the "panic" waves you emit when you see the accident coming. Many people overreact in such a situation, and end up flipping the vehicle, or panic and don't react fast enough. Such a system, coupled with lasers or infrared detectors that "see" the situation, could quickly make the right decision about how best to avoid the accident.

Understeering and oversteering, which we discussed earlier, are two more cases where cybernetics would be useful. They are dependent on the vehicle's weight distribution between the front and back axles. The cybernetic system could take over and apply the correct amount of steering.

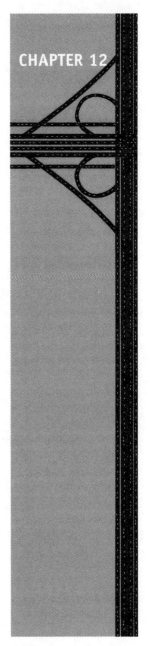

CHAPTER 12

Epilogue

The Final Flag

This concludes our tour of the physics of cars. I'm now reminded of a friendly argument I had with my dad many years ago, over which was the most mind boggling: the universe or the inner working of a car engine. As a physicist I was on the side of the universe, and as a mechanic and garage owner he was on the side of the engine. Most of what went on in the depths of space was foreign to him, but he found the miracle of the engine—how it turned over so many times per second and rarely broke down—to be truly amazing and fascinating. I'm not sure who won the argument; I think neither of us was able to convince the other. But I now understand his awe and reverence more than I did at the time. Trying to visualize all that is going on in an engine, and the speed with which it takes place, is difficult. The finely tuned coordination, the timing and intricacy of all those parts, is truly amazing. And the developments that are now going on make you wonder what the car of the future will be like.

We have looked at many different aspects of the car, with the emphasis being on the physics behind them.

We learned about the basic physics of driving, including concepts such as velocity, acceleration, momentum, and energy, and we calculated the force we experience when we take curves at various speeds. We determined how the weight of a car shifts when it is accelerated and braked, and we talked about the engine in considerable detail. The engine is, of course, what cars are really all about, so this discussion shouldn't have been a surprise. The measure of an engine is given by its torque and horsepower, and we looked into both of these concepts in detail.

We also looked at the electrical system and brakes. They don't generate as much enthusiasm from car fans as other parts of the car, but they are vital to the car's function. The time from 60 to 0 is certainly talked about a lot less than the time from 0 to 60, but it is just as important. After all, once the car is going, it's the brakes that get it to stop.

Suspension systems are, without a doubt, an integral part of the modern car, and whether most people realize it or not, it's one of the major features that sells the car when you take a test drive. A smooth, floating ride is as important to most people as gas mileage. We looked into what it is that gives us this smooth ride.

To most people the word *aerodynamics* conjures up an image of a sleek, highly streamlined airplane. But as we saw, the term is also important in relation to cars. Everything is tied up in the number known as the c_d, and since low c_d's improve gas mileage, it's natural to ask: How low is c_d likely to go? We now have a few cars down to .25 but an airplane wing has a c_d of .05, so there's little doubt that it will go lower. How much? I would not care to hazard a guess, but I'm sure progress will be made, and it wouldn't surprise me to see c_d's under .2 in the next few years. Much of the pressure for lower c_d's will no doubt come from the need for better gas economy.

We also looked at collisions. Even though it's something we don't like to think about, cars do crash into one another. And with physics we can learn a lot about the collision and in many cases determine who was at fault, if it wasn't obvious. Of most importance, though, is what it takes to make a car safer, and physics is of considerable help in this regard.

A book on cars wouldn't be complete without a chapter on car racing. In this book I discussed some of the things a race driver should know: weight distribution, balance, weight transfer, and strategy. An experienced race driver knows all of these things instinctively, but it's interesting to see why they are so important.

In the latter part of the book I took a turn into a different arena: traffic and traffic congestion. With more and more cars on the road, traffic is becoming an increasingly interesting topic. After all, something eventually has to be done about congestion. What is perhaps most interesting here is that scientists are now applying some of the latest advances in physics—chaos theory and complexity—to solve the problem. They may seem like an unlikely merging, but they have been giving intriguing and useful answers.

The book ends with a chapter on cars of the future and the devices that go along with them. Engineers have been speculating for years about what the car of the future will be like. I still remember seeing pictures of futuristic cars in magazines when I was young. They always seemed to have had large fins, which we now know are very nonaerodynamic.

In a book I once read, the author said on the last page that he felt like the pilot of an ocean liner who was on his last voyage. He felt melancholy about having to leave the liner, believing he still had a lot to offer, but he knew that he had to leave sometime. In some ways I feel the same way. There's certainly a lot more that can be said about the physics of cars, but I hope I have made a good start.

Bibliography

Aird, Forbes. *Aerodynamics.* New York: HP Books, 1997.

Appleby, John. "The Electrochemical Engine for Vehicles." *Scientific American* 281, 1 (July 1999): 74.

Ashley, Steven. "Driving the Information Highway." *Scientific American* 285, 4 (October 2001): 52.

———. "A Low Pollution Engine Solution." *Scientific American* 284, 6 (June 2001): 91.

Asimov, Isaac. *The History of Physics.* New York: Walker, 1966.

Birch, Thomas. *Automotive Braking Systems.* New York: Delmar, 1999.

Chinitz, Wallace. "The Rotary Engine." *Scientific American* 220, 2 (February 1969): 52.

Coghlan, David. *Automotive Braking Systems.* Boston: Breton Publishing, 1980.

DeCicco, John, and Marc Ross. "Improving Automotive Efficiency." *Scientific American* 271, 6 (December 1994): 52.

Genta, Giancarlo. *Motor Vehicle Dynamics.* Singapore: World Scientific, 1997.

Husselbee, William. *Automotive Transmission Fundamentals.* Reston, Va.: Prentice Hall, 1980.

Norbye, Jan. *The Car and Its Wheels.* Blue Ridge Summit: Tab Books, 1980.

Parker, Barry. *Chaos in the Cosmos.* Cambridge, Mass.: Perseus, 2001.

Pulkrabek, Willard. *The Internal Combustion Engine.* Upper Saddle River, N.J.: Prentice Hall, 1997.

Remling, John. *Automotive Electricity.* New York: Wiley, 1987.

Rillings, James. "Automated Highways." *Scientific American* 277, 4 (October 1997): 80.

Rosen, Harold, and Deborah Castleman. "Flywheels in Hybrid Vehicles." *Scientific American* 277, 4 (October 1997): 75.

Santini, Al. *Automotive Electricity and Electronics.* New York: Delmar, 1997.

Scibor-Rylski, A. S. *Road Vehicle Aerodynamics.* London: Pentech Press, 1975.

Sperling, Daniel. "The Case for Electric Vehicles." *Scientific American* 275, 5 (November 1996): 54.

Wilson, S. S. "Sadi Carnot." *Scientific American* 245, 2 (August 1981): 134.

Wise, David, ed. *Encyclopedia of Automobiles.* Edison, N.J.: Chartwell Books, 2000.

Wouk, Victor. "Hybrid Electric Vehicles." *Scientific American* 277, 4 (October 1997): 70.

Zetsche, Dieter. "The Automobile: Clean and Customized." *Scientific American* 273, 3 (September 1995): 102.

Internet Sites

www.members.home.net/rck.html
www.aerodyn.org
www.theatlantic.com/issues/2000
www.sciam.com/explorations
www.gallery.uunet.be/heremanss

Magazines

Motor Trend
Automobile
Road and Track

Index